MW00466968

ISBN 978-1-332-73400-9
PIBN 10294615

Practical Lessons

ON

The Lever Escapement Its Tests, Errors Their Detection and Correction

By T. J. WILKINSON

Author of "Practical Studies in the Lever Escapement," "Train, Wheel and Pinion Problems"; "Length of Lever and Roller Jewel Radius," etc., etc.

TECHNICAL PUBLISHING COMPANY
2258 North Front Street
Philadelphia, Penna.

This book on the Lever Escapement is sent forth with the ope that students in horology, the young and the advanced, will nd pleasure and profit in using it.

A part of this volume recently appeared in the columns of the *ewelers' Circular*, to the Editors of which the author desires to eturn thanks for favors extended.

The system outlined and described in the Lessons has had a horough and successful test, not only with beginners, but with orkmen of wide experience, and will be found adapted to the eeds of both.

Upon the completion of our last escapement series, many ritten commendations from unknown readers were received. he following is an extract from one of these letters:

"I am a practical watchmaker for the past twenty years, and, lthough considered a skilled workman, I have profited more from our recent publication in the *Jewelers' Circular than from any mount of practice.*"

This plain statement should spur others to add to the knowl-dge they now possess, increased skill being an invaluable asset.

To tell a man you know the road to "Wheeltown" and then dmit you are not capable of directing him on his way, is and as been the position of many as regards the Lever Escapement nd its tests. A want of nomenclature has been largely respon-ible for this. The writer therefore at different times coined escriptive terms, such as "Angular Test," "Corner Test," "Cor-er Safety Test," etc. The term "Tripping," an old one in horol-gy, we have divided up into three parts, giving to each a def-nite meaning, namely, "Corner Trip," "Guard Trip," "Curve rip." The adoption of exact expressions renders explanation and nstruction more profitable for all.

We plead guilty to repetition in the Lessons. Some one has isely remarked, "Reiteration is education," and our experience as proven its truthfulness. No further comment on this score s necessary.

The methods outlined in the book provide students with an ndoubted short cut to Escapement knowledge, both practical nd theoretical. Previously, command over the Escapement and s problems was acquired by years of bench practice and experi-ment, which, because undirected, often failed of its purpose or ead to erroneous decisions. Such drawbacks the book's teachings vert. If the reader will study the Lessons, work out the solu-ion of each Test Lesson and make use of the Questions, quickly is work will show a decided improvement, alike pleasing and atisfactory to all.

Philadelphia, 1916. T. J. WILKINSON.

LESSON 1

ADVICE—PRACTICE

1. We shall devote this our opening lesson to two things—advising students how to gain most profit from this study of the lever escapement and to describing in brief elementary terms such horological words as are necessary to our subject. Students are asked to study each item as a link in the chain of escapement knowledge. As you read through the lessons mark, learn and thoroughly digest every paragraph. In a work of this class repetition is necessary. Repetition is a good method of teaching. It impresses facts important to be known and remembered; hence we have not hesitated to employ it.

Systematic study and work will save young watchmakers years of unsatisfactory experience at the bench. The results to be obtained are, therefore, worthy of the effort. A certain amount of theory is necessary to understand the basis of escapement form and construction. The extent of theory in the following lessons is limited to such as can be practically made use of. The escapement tests are all practical; you must be acquainted with them if you desire to do your work well and quickly. In connection with them we advise experimental work; prove all things mentioned in the following lessons; but above all, see that you understand each and every move. Depart from the ways of the average watchmaker, who when asked to explain various features of the escapement or errors connected with same replies by saying, "I know, but I can't explain"; yet the same man will give you a clear, intelligent description as to how he cleans a watch. He knows the latter, but the escapement is somewhat of a muddle and puzzle to him.

The student who will carefully study the following lessons will know and will be able to explain the whys and wherefores of escapement problems such as he encounters every day at the bench.

REQUIREMENTS FOR PRACTICE

2. To combine study with practical work students should provide themselves with a 16-size three-quarter plate, or else a bridge movement of either Waltham, Elgin, Illinois, Rockford or Hamilton type. The 16-size three-quarter plate movement is favored because the parts are fairly visible; besides, examinations and tests are more easily conducted, the parts being more accessible. For

experimental purposes it is also necessary that students obtain several old movements, including some New York Standards. For the latter an extra supply of levers and table rollers is desirable. As the movements named possess composition levers, they are well adapted for research escapement work. Students when sufficiently advanced should make a point of examining all escapements. By this means they become familiar with the routine of escapement testing.

In the experimental work various errors should be created; much can be learned along this line. For instance, by bending the lever we can throw the escapement "out of angle," or we can create this defect by making the drop lock on one pallet stone greater or less in amount than that found on the opposite pallet.

Composition levers are easily altered, either lengthened or shortened, as desired. The result of changing the lever's length can be shown and detected by the tests. Therefore, when experimenting we can, by employing the tests, determine how and in what manner an escapement is deranged and what alterations are necessary to restore the escapement to a normal condition.

Experimental research work will be found the best and surest road toward a complete mastery of the problems of the lever escapement. In a practical way—namely, at the bench—we hope our readers will apply the methods about to be set forth.

LESSON 2

DEFINITIONS

BALANCE

3. *Balance.*—The vibratory wheel of a watch, which, in conjunction with the hairspring, controls the progress of the hands.

4. *Balance Arc.*—For definition of balance arc see No. 106.

5. *Balance Arc of Vibration.*—See No. 108.

6. *Balance, Supplementary Arc.*—See No. 107.

7. *Balance Staff.*—The axis of the balance.

8. *Balance Spring.*—The fine coiled spring attached by a collet to the balance staff. Frequently it is termed the hairspring. This spring assists the balance to vibrate.

9. *Balance Wheel.*—The escape wheel of a verge watch. It is incorrect to employ the term "balance wheel" to express the balance as used in the lever and other watches.

BANKING

10. *Banking.*—In a lever watch the term "banking" implies that the roller jewel, due to an excessive vibration of the balance, strikes on the *outside* of the lever horn. This error might result from using a mainspring of too great a strength.

11. *Banking Pins.*—In modern American watches the banking pins placed on each side of the lever are eccentric extensions from the screws. This method of construction allows of a great deal of latitude in controlling the amount of lock and the lever's motion.

12. *Bank.*—A shortening of the term "banking pin."

13. *Banked to Drop.*—An escapement term implying that slide or second lock has been eliminated. To bank an escapement to drop, it is necessary to close in the banking pins to such an extent that drop lock *only* is present.

14. *Banked to Drop, Elgin Type.*—When an escapement of the Elgin type is banked to drop, *a slight freedom exists* between guard point and edge of table (guard freedom); also between slot corners and roller jewels (corner freedom).

15. *Banked to Drop, South Bend Type.*—When an escapement of the South Bend type is banked to drop, *no* freedom is found either of the guard point with edge of table or of the roller jewel with the slot corners.

16. *Banked for Slide.*—This expression refers to the presence in an escapement of the slide or increase of the lock which (when the bankings are open) follows drop lock.

DROP AND SHAKE

17. *Drop.*—When a tooth of the escape wheel is discharged from either pallet stone the escape wheel is immediately released from all contact. The wheel's motion is then entirely free. This free flight of the wheel, termed "drop," ceases the moment another tooth meets the locking face of an intercepting pallet jewel. Drop is also defined as the space through which an escape wheel moves without doing any work.

18. *Inside Drop.*—Is that space through which the escape wheel moves whenever a tooth leaves the releasing corner of the entering pallet jewel. Inside drop ceases the moment another tooth strikes the locking face of the exit pallet.

19. *Outside Drop.*—Is that space through which the escape wheel moves when a tooth becomes disengaged from the releasing corner of the exit pallet. Outside drop ceases the moment another tooth meets the locking face of the entering pallet.

20. *Shake.*—The term "shake" implies that position of the pallet jewels with the adjacent teeth of the escape wheel where least freedom of parts is found to exist. The following is more explanatory:

21. *Inside Shake.*—The position of least freedom of the escape wheel teeth embraced *between* the pallet jewels at the moment of unlocking. A practical observation by the student will make this clear. Bring the tooth at rest on the locking face of the exit or discharging pallet down to the pallet's lowest locking corner; then note the space separating the back of the entering pallet from the heel of the tooth just behind it. The space seen separating the point of the tooth from the back of the entering pallet represents the amount of inside shake. When in this position the parts have least freedom.

22. *Outside Shake.*—This term refers to that position of the escape wheel teeth *outside* the pallet jewels where least freedom exists. Practical demonstration will make this clear. To discover the amount of outside shake bring the tooth of the escape wheel found resting on the locking face of the entering pallet jewel down to the lowest locking corner of this stone. Hold the parts in this position and notice the space separating the back of the exit pallet from the heel of the tooth just behind it. The space observed represents the outside shake and shows the least freedom of the parts, which occurs always at the moment of unlocking.

DRAW

23. *Draw.*—The force which holds the lever against its bank. Draw is chiefly the result of the angle given to the pallet jewel's

locking face. The force of draw is also helped by the angle on the locking faces of the teeth of the escape wheel. It is advisable for students to learn and compare the terms "draw," "run" and "slide."

24. *Draw* Lock.—See slide.

ESCAPEMENT

25. *Escapement.*—That part of a lever watch which changes the circular force of the escape wheel into the vibratory motion of the balance.

26. *Single Roller Escapement.*—A single roller escapement, as the name implies, possesses but one table roller. The office of this roller is as follows: First, to carry the roller jewel; second, its periphery or edge is a part of the safety action; third, the crescent or passing hollow provides space for the free passage of the guard point.

27. *Double Roller Escapement.*—Escapements of this type possess two rollers. The larger, or impulse table, carries the roller jewel. The smaller, or safety roller, is an important factor in the safety action. It also contains the crescent or passing hollow.

28. *Right-Angled Escapement.*—In a right-angle escapement we find the line of centers of pallet and balance crossed at right angles by the line of the escape wheel.

29. *Straight-Line Escapement.*—A straight-line escapement is one wherein we find the pallets, lever and balance all planted in a straight line.

ESCAPE WHEEL

30. *Club Tooth Wheel.*—This term describes the shape of an escape wheel tooth as used in American watches.

31. *Ratchet Tooth Wheel.*—A wedge-shaped form of tooth used in escape wheels of English-made watches.

32. *Lift on Tooth.*—The slant on the upper face of a club tooth. In Fig. 1 the line A B embraces "the lift."

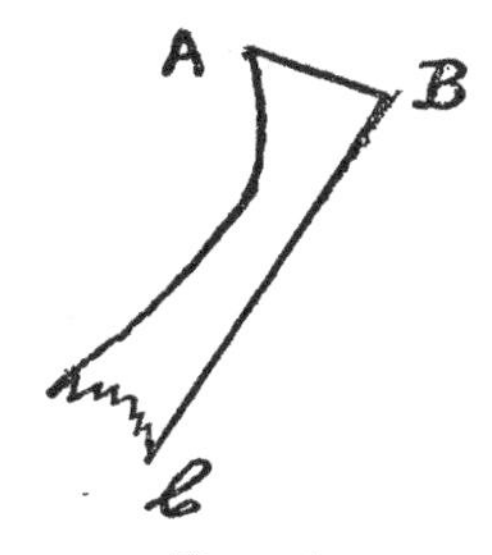

FIG. 1

33. *Pitch of Tooth's Locking Face.*—The angle found on the back of each tooth. (See Fig. 1, B to C.)

34. *Tooth's Impulse Face.*—See "Lift on Tooth."

35. *Heel of Tooth.*—In the illustration Fig. 1 the point A is the heel.

36. *Toe of Tooth.*—The point B (Fig. 1) is the toe of the tooth.

LOCK

37. *Locking.*—The overlapping contact of an escape wheel tooth on a pallet jewel's locking face.

38. *Drop Lock.*—That point of contact of a tooth of the escape wheel as it drops onto the pallet jewel's locking face. Drop lock is also termed the first, or primary, lock.

39. *Drop Lock, Elgin type.*—Of the total lock in an Elgin type of escapement, about two-thirds represents drop or first lock.

40. *Drop Lock, South Bend Type.*—One-half of the total lock in this type of escapement will equal the amount of drop lock.

41. *Slide or Slide Lock.*—This expression refers to the further increase of the lock following the drop lock. Slide lock is variously known as draw lock, second lock or simply as slide.

42. *Slide, Elgin Type.*—About one-third of the total lock in an Elgin type of escapement should be slide.

43. *Slide, South Bend Type.*—One-half of the total lock in an escapement of this type will be slide or second lock.

44. *Remaining or Safety Lock.*—The amount of lock of the tooth on the pallet jewel's locking face which remains when the safety tests are used. (See corner safety and guard safety tests.)

45. *Total Lock.*—A term including both drop and slide lock. Total lock is therefore the sum of both.

OVERBANKING

46. *Overbanked.*—The term "overbanked" expresses the fact that the roller jewel has come to rest against the outside of the lever horns. Whenever, owing to some defect, the lever passes in an irregular manner over to the opposite bank, the roller jewel then strikes the outside of the lever and the watch is said to be overbanked.

LEVER

47. *Lever.*—The flat metal bar which conveys and transmits motion to the balance. To the lever bar is attached the pallet arms. The end of the lever associated with the roller jewel is termed the fork.

48. *Acting Length of Lever.*—The distance from the pallet center to the slot corners represents the lever's acting length.

49. *Lever Horns.*—The circular sides of the fork leading into the slot.

50. *Lever Slot or Notch.*—The slot cut into the lever bar, centrally located below and between the lever horns.

51. *Fork.*—A term including both slot and horns of lever.

52. *Run of* Lever.—The continued motion of the lever toward its bank which takes place when slide or second lock is present. The amount of run always equals the amount of slide.

PALLETS

53. *The* Pallets.—The metal body attached to or a part of the lever. This includes the pallet arms and jewels. Together the pallet jewels and metal body comprise the pallets. By means of these parts the escape wheel transfers its energy to the lever and balance.

54. Pallet *Arms.*—The metal body which contains the pallet jewels.

55. Pallet *Jewels.*—The jewels or stones inserted in the pallet arms for the purpose of receiving and transmitting the energy delivered by the escape wheel.

56. Pallet *Staff.*—The axis of the pallets and to which the pallet arms and lever are attached.

57. *Entering or Receiving* Pallet.—That pallet jewel over which a tooth of the escape wheel slides in order to enter between the pallet stones.

58. *Exit or Discharging* Pallet.—That pallet stone which an escape wheel tooth slides over in order to make its exit outside the pallet jewels.

59. Pallet's *Impulse Face or* Plane.—The lower surface of a pallet jewel upon which the escape wheel teeth act.

60. *Lift on* Pallet.—The pitch or slant of the impulse plane.

61. Pallet's Locking *Face.*—That portion of a pallet jewel upon which a tooth of the escape wheel drops and locks.

62. *Releasing Corner of* Pallet.—That point on a pallet stone's impulse face where the tooth is released from contact with the pallet.

63. Pallets *Circular.*—A type of pallet so constructed that (central) points located on each pallet jewel's *impulse* face, *midway* between the entering and releasing corner, are exactly at the same distance from the pallet center. This arrangement, of course, places the *locking faces* at an *unequal* distance from the center of the pallet. Circular pallets are used in American-made watches.

64. Pallets *Equidistant.*—Pallets that have their *locking faces equally distant* from the pallet center are known as equidistant pallets. This form is found only in foreign-made watches of very high grade.

ROLLER JEWEL

65. *Roller-Jewel.*—The long, thin cylindrical-shaped jewel inserted in the table roller; also called the "jewel pin," the "impulse pin."

66. *Roller Jewel Radius.*—The distance from the center of the roller to the face of the roller jewel.

SAFETY ACTION

67. *The Safety Actions.*—In a single roller escapement the safety actions include the following: Guard pin, edge of roller, roller jewel, corners of lever slot and a small portion of the horns close to the slot corners; and we may include the drop locks. In a double roller escapement the parts are the guard finger, the roller jewel, the lever horns, the slot corners, safety roller and the drop locks.

68. *Guard Pin.*—The upright pin inserted in the lever bar just behind the slot. The term "guard pin" applies to single roller escapements.

69. *Guard Finger.*—The pin extending from the lever bar of a double roller escapement and pointing toward the safety roller.

70. *Guard Point.*—As used in these lessons, this term expresses either the guard pin or the guard finger.

71. *Guard Radius.*—The distance from the center of the pallet staff to the outside of the guard point.

72. *Guard Test and Guard Safety Test.*—For definitions consult Nos. 82 and 85.

ROLLERS

73. *Table Roller or Impulse Roller.*—The circular disc attached to the balance staff and into which the roller jewel is inserted. In single roller escapements the crescent or passing hollow is cut into the edge of the roller.

74. *Safety Roller or Safety Table.*—The smaller sized disc found in double roller escapements. In its edge is located the deep, wide cutting called the crescent or passing hollow.

75. *Crescent or Passing Hollow.*—The circular cut or bight formed in the edge of a roller. Its purpose is to provide for the necessary intersection of the guard point during the latter's excursion from bank to bank. Width and depth are important features of the crescent.

76. *Diameter of Roller.*—The diameter of a roller is a line drawn from any point on its edge to a point directly opposite, but the line so drawn must pass through the centre of the roller.

TESTS AND TEST TERMS

77. *Angular Test.*—A test used to show the relationship existing between the amount of drop lock and the lever acting length. Assuming the drop locks as correct, the angular test will show if the lever's length is correct, long or short. This test is also employed to show if an escapement is "in angle" or "out of angle."

78. *Corner Test.*—A test used to discover the relation of the roller jewel with the slot corners of the lever. It is most accurate when made under banked-to-drop conditions.

79. *Corner Freedom.*—The freedom found by the corner test between the roller jewel and the slot corners.

80. *Corner Safety Test.*—That subdivision of the corner test which shows if the remaining or safety lock is present or absent.

81. *Curve Test.*—The test used to discover if the curves of the lever horns are correctly related to the roller jewel. It is mostly applied to double roller escapements.

82. *Curve Safety Test.*—A subdivision of the curve test whereby we learn about the condition of the remaining or safety lock.

83. *Guard Test.*—This test is employed to determine the position of the guard point with reference to the edge of the roller.

84. *Guard Freedom.*—The freedom found between the guard point and edge of roller.

85. *Guard Safety Test.*—A subdivision of the guard test which enables us to learn about the condition of the remaining or safety lock.

TRIPPING

86. *Tripping.*—Tripping is the irregular act of an *escape wheel tooth entering* on to a pallet jewel's *impulse face* owing to a fault in the safety action. When the safety tests are employed tripping is shown by the *absence* of a safety or remaining lock.

87. *Corner Trip.*—The want of a remaining or safety lock discovered by means of the *Corner Safety Test.*

88. *Curve Trip.*—A lack of safety lock developing when the *Curve Safety Test* is applied.

89. *Guard Trip.*—An absence of safety lock found by means of the *Guard Safety Test.*

LINE OF CENTER—OUT OF ANGLE—ADJUSTING LET-OFF

90. *Line of Centers.*—The line of centers in the lever escapement is a line drawn from the center of the pallet hole jewel to the center of the balance hole jewel. The lever as it travels from bank to bank moves an equal distance on each side of this line.

91. *Out of angle.*—When an escapement is out of angle *and the watch banked to drop, the lever,* as judged by the line of centers, moves an *unequal* distance on each side of this line in order to reach its bank. The guard point indicates an escapement as out of angle when it has *greater freedom* on one side of the roller than on the opposite side. The roller jewel shows the escapement as out of angle by *an inequality of freedom* with

each slot corner. Out of angle is corrected by altering the pallet stones or bending the lever, or perhaps both. To correctly decide if an escapement is out of angle *it must be banked to drop*.

92. *Let-off Adjusting.*—Adjusting the "let off" is a factory expression denoting the escapement is out of angle. It implies that one or both pallet stones need altering or the lever requires bending to provide the guard point with an equal amount of shake on each side of the roller and to give the roller jewel an equal amount of freedom with each slot corner. As adjustments are made with the escapement banked to drop, the final result is an *equal* motion of the lever on each side of its line of centers.

LESSON 3

ANGLES—CIRCLES—DEGREES

NECESSARY KNOWLEDGE

93. Every student aspiring to become master of the principles of escapement construction must at least possess a practical working knowledge of angles, degrees, etc. A knowledge of escapement drafting is also desirable. As a majority of students lack this necessary instruction, we have in this and succeeding chapters enumerated such points as have a direct and practical bearing on our subject.

ANGLES

94. An angle is the *opening* between two lines which meet at a point. The meeting place is termed the *vertex;* the lines defining the angle are called its *sides.* When an angle stands alone it can be named by the letter placed at the meeting point of the lines. For instance, we refer to angle B (Fig. 2). Should two

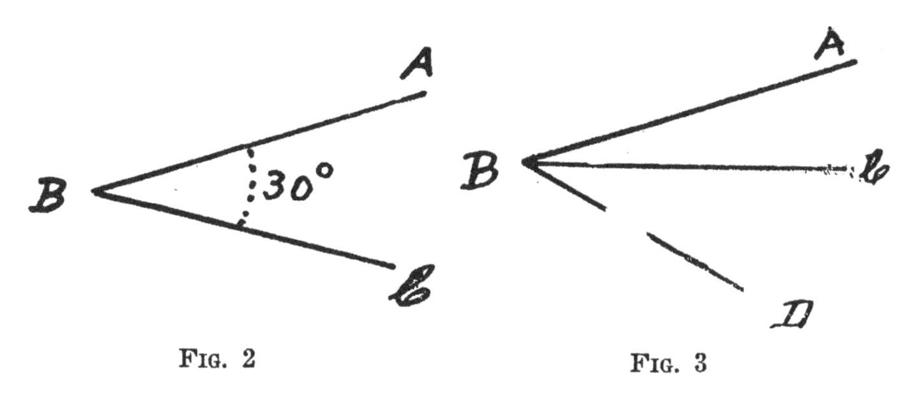

Fig. 2 Fig. 3

or more angles meet at a common center they are named by three letters, as A B C or C B D (Fig. 3). Angles are spoken of as so many degrees in width. The greater the divergence of the lines of an angle the greater the number of degrees contained by that angle. The *size* of an angle is measured by the extent of its opening relative to the 360 degrees in a circle. We shall therefore discuss angles and their degrees in our discussion of circles.

CIRCLES

95. A circle is a plane figure bounded by a curved line, called its *circumference.* All lines drawn from the center of the circle to its circumference are equal in length. The lines A E, E C, E D and E B (Fig. 4) are equal. The diameter of a circle is any

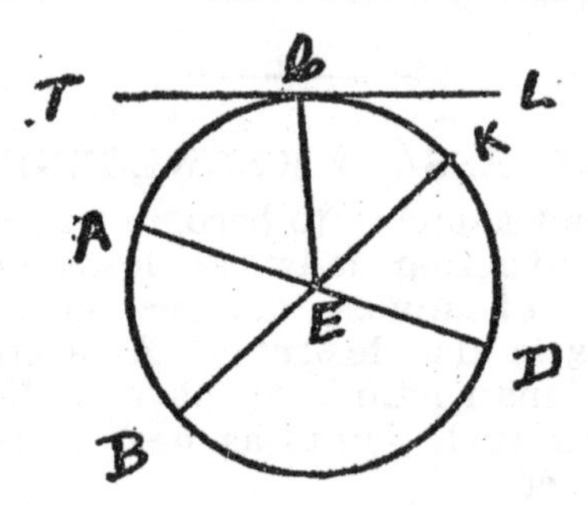

FIG. 4

straight line drawn from circumference to circumference and passing through the circle's center. The lines A D and K B are diameters of the circle (Fig. 4).

RADIUS

96. The radius of a circle is any straight line *drawn from the center to the circumference.* The lines E C, E B, etc., are radii of the circle (Fig. 4). ("Radius" is the singular; "radii" the plural.)

ARC

97. An arc of a circle may be any portion of the circle's circumference. Thus A to C (Fig. 4) is an arc.

TANGENT

98. A tangent to a circle is a straight line which touches the circle at only one point. The tangent is always perpendicular to the radius drawn to that point. In Fig. 4 the line T L is tangent to the circle at the point C and the radius E C is the tangent's perpendicular; consequently the angles T C E and L C E contain 90 degrees each.

SEMICIRCLE

99. A semicircle, as the name implies, is one-half of a circle; every semicircle contains 180 degrees.

QUADRANT

100. A quadrant is the one-fourth part of a circle, and contains 90 degrees.

RULES APPLYING TO CIRCLES

101. Radius equals one-half the diameter.
Radius multiplied by 2 equals one diameter.
Radius multiplied by 2 by 3.1416 equals circumference.
Diameter multiplied by 3.1416 equals circumference.

DEGREES

102. The circumference of every circle is supposed to be divided into 360 equal parts; each division or part is termed a degree. The degree is again subdivided into minutes;-each degree contains 60 minutes.

TABLE OF SIGNS

103. Three hundred and sixty degrees is equal to one circle.
Sixty minutes is equal to one degree.
Degree sign (°) used for degrees.
Minute sign (′) used for minutes.

LENGTH OF ONE DEGREE

104. The length of one degree changes with the size of the circle of which it is the 1/360 part. The length of one degree on the earth surface is about 60 geographical miles. The size of one degree on the circumference of a circle measuring 360 feet would equal one foot ($360 \div 360 = 1$). From the foregoing statements we learn that the size of one degree is a varying factor altogether dependent on the size of the circle. The size of a degree in watch-work likwise varies with the size of the circle of which it is the 1-360 part. For instance, two degrees of lock in an 18-size escapement will measure more than two degrees of lock in an 0-size watch. Again, two degrees of lock measures more on the entering pallet of any American watch than two degrees do on the exit pallet of the same watch, because, being circular pallets, their respective locking faces are placed at different distances from the pallet center. This the student can demonstrate by measuring the distance from center of pallet staff to lowest locking corner of each pallet jewel as directed in item No. 180.

PROTRACTORS OR ANGLE MEASURES

105. Students can quickly advance their practical knowledge of angles and degrees by becoming acquainted with an instrument for the measurement of angles called a protractor, as illustrated

5. Suppose we wish to measure the size of angle B (Fig.
can do so by placing the center 0 of the protractor at B,
egree mark extending along the line B A. When so placed

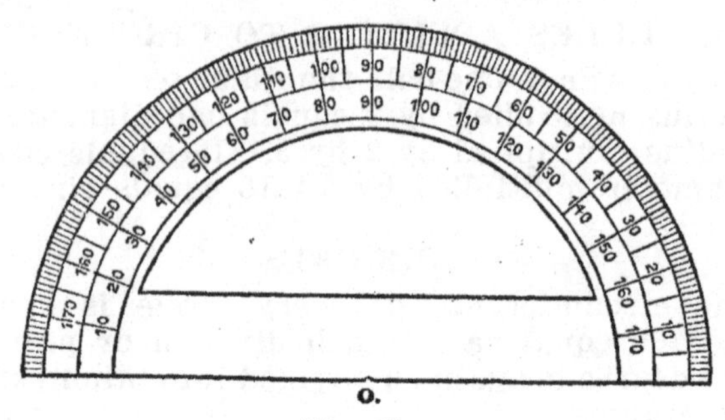

FIG. 5

easy matter to count off the degrees enclosed by A B C as
le of 30 degrees. It is not a difficult matter to become
r with the various methods of using the protractor. Cheap
tors are usually unreliable, and the student who intends
ue a thorough course in escapement drafting is advised to
se a reliable instrument.

LESSON 4

BALANCE ARC—SUPPLEMENTARY ARC—ARC OF VIBRATION

106. *Balance Arc.*—The definition of an arc, as before given, is any part of a circle's circumference. The *balance arc* is that part of the arc of vibration during which *the roller jewel is in contact with the lever.* The extent of this balance arc varies; usually it is around 30 to 40 degrees, smaller arcs being employed in double than in single roller escapements. The extent of the balance arc is, of course, measured from the balance center.

107. *Supplementary Arc.*—The *supplementary arc* represents that portion of the arc of vibration of the balance during which the roller jewel is detached from the lever; we might also because of its detachment, term it the "free arc."

108. *Arc of Vibration.*—The arc of vibration equals the sum of supplementary arc plus the balance arc. Accordingly the arc of vibration represents the full swing or motion of the balance.

109. *Motion of Balance.*—What constitutes a good motion is a question of dispute. Indeed, many watchmakers have but little conception of what the proper motion of a watch should be. To determine the arc of vibration takes a combined trained eye and mind; these every watchmaker should cultivate. It has been demonstrated that when a balance gives one and one-eighth to one and one-quarter turns it *neutralizes* slight inaccuracies in the poise of the balance. One and one-eighth turns means an arc of vibration of 405 degrees; one and one-quarter turns expresses an arc of 450 degrees; one and one-half turns equals a vibratory arc of 540 degrees. Regarding the question of a good motion, the man at the bench usually prefers an arc of 450 to 540 degrees. The vibratory arc, however, should not exceed 540 degrees, else there is danger of developing a banking error. An examination of the highest grade watches will reveal the fact that arcs of 450 degrees are most favored.

LESSON 5

PALLETS—CIRCULAR, EQUIDISTANT AND DRAFTING

110. *Form of Pallet Jewel.*—The acting faces of a pallet jewel are A B and A C (see Fig. 6). The part A B is the jewel's locking face. Upon some part of its surface the escape wheel tooth drops

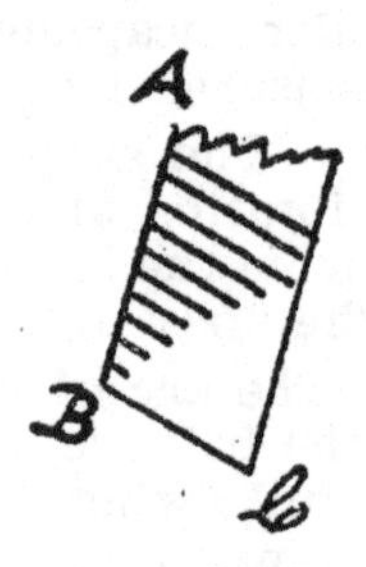

FIG. 6

and locks. Notice the slant of A B away from the point B. A B is shown as an inclined plane. The purpose of this slant is to help draw the pallet jewel deeper into the escape wheel and to hold the lever against its bank. The force which produces the effects just mentioned is termed "draw." That part of the pallet from B to C is termed the lifting or impulse face of the pallet jewel. The impulse face assists in moving the lever from bank to bank. It is directly associated with the lifting face found on a tooth of the escape wheel. The combined lifting or impulse faces, located on tooth and pallet, are directly responsible for the blow given to the roller jewel by the slot in the fork.

111. *Angles Shaping the Pallet Jewel.*—The angles which control the shape of a pallet jewel are three in number. These three angles arise from three different points, namely, the pallet center at D, the escape wheel center at E and the pallet corner at B (see Fig. 7).

112. *Angle of Impulse.*—The angle H D K (Fig. 7) arises at the pallet center D. This angle is known as the impulse or lifting

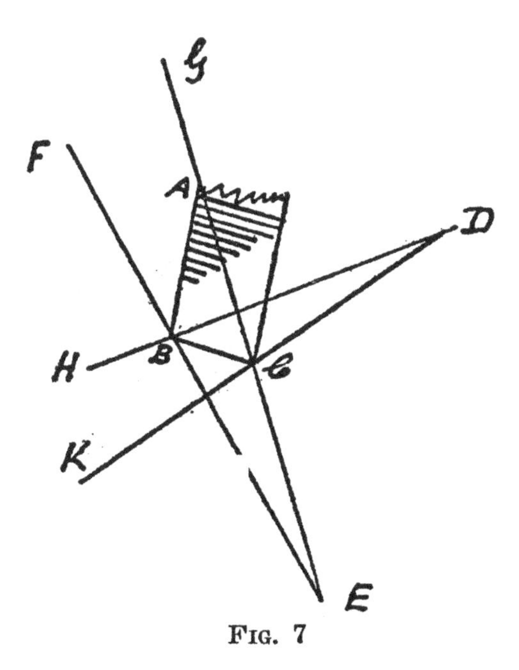

FIG. 7

angle. The impulse angle in conjunction with the angle of width, viz., F E G (Fig. 7), defines B C, the pallet's impulse face.

113. *Angle of Width.*—The width of a pallet jewel is governed by the size of the angle F E G, the point of origin of this angle being the escape wheel center E (Fig. 7).

114. *Draft Angle.*—The draft angle F B A (Fig. 7) starts from the point B. This point is located at the lowest locking corner of the pallet jewel. The degrees of slant of the line A B are reckoned as away from the line F B.

115. *Angle of Lock.*—The angle of lock of tooth on the pallet jewel is not represented in the drawing (Fig. 7) for the reason that the angle of lock has nothing to do with the shape of the pallet stone.

116. *Circular Pallets.*—The type of pallet used in American watches is known as the "circular pallet." The locking face of the entering pallet jewel is further from the pallet center than is the locking face of the exit pallet. That part of each pallet stone, situated midway betwixt the entering and discharging corners of the pallet's impulse plane, is at an equal distance from the pallet center. Pallets of the circular type are easily recognized by means of a depthing tool. The V-shaped end of the sliding rod

of the depthing tool is placed over the pivot of the pallet staff and the tool adjusted so that the sharp point of the other rod touches the *locking corner* of the *receiving* pallet. It will then be found, on swinging the point of the tool over the opposite pallet, that the tool's point just touches the *discharging corner* of the *exit* pallet. If we adjust the point so it stands centrally over one pallet it will be found that the point also stands centrally over the opposite pallet. In this manner we can prove if pallets are of the circular type.

117. *Equidistant Pallets.*—Equidistant pallets are found only in the higher grades of foreign-made watches. They are easily recognized by using a depthing tool in the manner previously described. We will, however, find this difference, that with the point adjusted to touch the lowest *locking corner* of the *entering* pallet, and then swung over on to the exit pallet, the point of the tool will be found to touch the *locking corner* of the *exit* pallet. This experiment will prove that the locking faces are equidistant from the pallet center; therefore the term "Equidistant Pallets" is used to describe them.

118. *Drafting Circular Pallets.*—We shall now briefly explain the drafting of a circular pallet. It is not the purpose of these lessons to delve extensively into the subject of drafting. If we were to attempt it an entire volume would have to be devoted to the subject. Besides, from the writer's experience, personal instruction is necessary if the student is to greatly profit from the making of complete escapement drawings. Our present purpose is to supply the student with the simplest theoretical explanation of the various escapement parts. This will be sufficient to insure a foundation for further advancement in this interesting and useful branch of educational horology.

119. *Specifications.*—Distance of center of receiving pallet to center of exit pallet, 60 degrees; width of pallets, 6 degrees; lift on pallets, 5½degrees; total lock, 2 degrees. Of the total lock 1½ degrees is drop lock, the remaining one-half degree being slide. Draft angle on pallet, 12 degrees. Commence by drawing the line B H (Fig. 8); the point B represents the escape wheel center. Somewhere along the line B H the pallet center will be located. Its location will be later determined. With B as center, describe the arc C C. On each side of the line H B lay off two angles, each containing 30 degrees. The angles D B H and H B E each contain the required 30 degrees. Draw the lines F A and A G; these lines intercept D B and E B exactly at right angles. The lines F A and A G are tangents to the circle C C and meet at the point A on the line B H. Their meeting point is at A, the pallet center. Our specifications call for a pallet width of 6 degrees. As we are dealing with circular pallets, we accordingly lay off one angle of 3 degrees to the left of the line D B and

FIG. 8

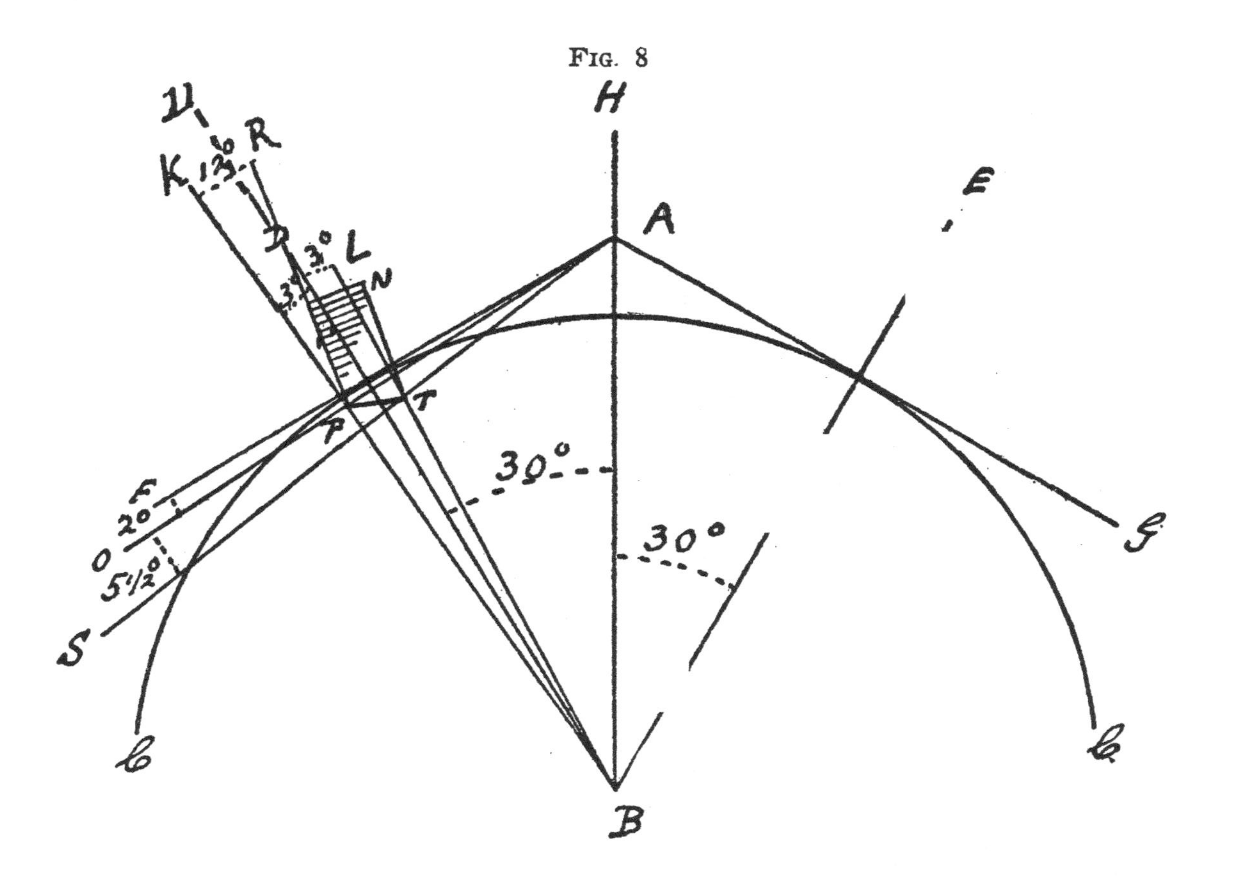

another angle of equal size to the right of D B. The angle K B L is therefore one of 6 degrees, as called for by our specifications. The angle of lock is to equal 2 degrees, therefore below the tangent line F A we lay out the angle F A O, containing 2 degrees. At the point where the line O A intercepts the line K B we locate the corner of the pallet jewel (see P in drawing). From P, and away from the line P K, we lay off an angle of 12 degrees, as shown by K P R. This is the draft angle of the pallet, and accordingly R P is the pallet jewel's locking face. Below the line O A lay off an angle of 5½ degrees. The angle O A S is the required angle. This is the angle of lift for the pallet stone. Observe where the line S A and B L cut each other; this point we have marked T. Connecting the point P with the point T gives us the line P T. This line marks out the lifting or impulse face of the pallet jewel. The back of the pallet jewel N T is simply drawn parallel to the locking face R P, which completes the drawing.

LESSON 6

THE ESCAPE WHEEL—DRAFTING

120. *Escape Wheel Teeth.*—The acting parts of a club tooth are two in number, namely, the lifting plane A B (Fig. 9) and the incline B C which starts at the locking corner B. Of the whole line B C practically the corner B alone comes into action, this

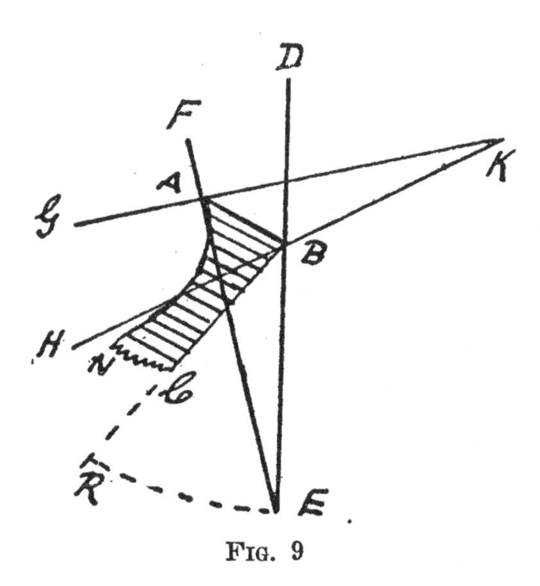

Fig. 9

being the part which rests on the pallet jewel's locking face. The line A N takes no part in the escapement action, the undercutting from the point A being for clearance between pallet and tooth. This statement, like others, should be verified by the student from actual observation of an escapement in action.

121. *Tooth's Impulse Face.*—The lifting face or impulse plane of a tooth is defined by the line A B (Fig. 9). The impulse face on the tooth combined with the impulse plane on the pallet jewel imparts, through the medium of the fork, the force necessary to keep the balance vibrating.

122. *Tooth's Draft Angle.*—The line B C (Fig. 9) of the tooth is given a very decided slant in order that only the corner B will rest against the locking face of the pallet jewel. The effect is, that friction between tooth and pallet is lessened. This reduction of friction greatly improves the "draw." The purpose of draw is to retain the lever against its bank.

123. *Angles Shaping a Tooth.*—The angles which give shape to a club tooth are three in number. They are the angle of width which defines the width of the tooth, the angle of impulse which governs the amount of lift and the angle which gives form to the slant on the back of the teeth. These angles arise from three different points.

124. *Angle of Impulse.*—The angle controlling the amount of impulse or lift on a club tooth originates at K (Fig. 9), the point K being the pallet center. This angle is illustrated as enclosed between the lines G K H.

125. *Tooth's Angle of Width.*—The width of a tooth of the escape wheel is governed by the angle F E D (Fig. 9). The starting point of this angle is at E, the escape wheel center.

126. *Tooth's Draft Angle.*—The draft angle, or angle which defines the slant B C, is shown as enclosed by the lines E B R. The starting point of this angle is at B, the degrees of slant being reckoned away from B E. Should we desire to make a complete drawing of an escape wheel, a number of radial lines should be drawn from the locking corner of each tooth to the center of the escape wheel. The degrees of slant for B C would be counted from each line as illustrated by the angle R B E (Fig. 9).

127. *The Angle of Drop.*—The angle of drop arises at the escape wheel center. The amount of the angle of drop in the better grade watches is about 1½ degrees. In watches of poorer construction it is often greater and frequently irregular. A practical method for estimating the angle of drop is given in Lesson 13.

128. *Drop.*—When a tooth of the escape wheel becomes detached from the releasing corner of either pallet jewel a free flight of the escape wheel through space results. This free motion of the escape wheel is termed its drop. Drop commences the moment a tooth separates itself from a pallet jewel and ceases the instant another tooth is caught on the intercepting locking face of the opposing pallet. We can define "drop" as the space through which an escape wheel moves without doing any work.

129. *Drop and Shake.*—The angle or amount of drop visible in any escapement does not represent *the least* freedom between the pallet jewels and the teeth of the escape wheel. The position wherein the least freedom exists between the teeth and pallets is spoken of as their "shake." The student should experimentally

determine the shake or position of least freedom in the following way: Bring a tooth down on to the lowest locking corner of a pallet jewel and note the space separating the back of the opposite pallet from the heel of the tooth just behind it. A brief examination will prove that the space so seen, viz., the shake, is less than its corresponding drop. The amount of drop and shake present in any escapement can in a practical way be estimated by using the pallet's width as a standard, as explained in Lesson 13, paragraphs 184 and 184A.

130. *Shake.*—The amount of shake present in any escapement is always closely related to the angle of drop, because shake equals the drop *minus* the recoil of the escape wheel. Like drop, we have two classes of shake, namely, outside and inside shake. Whenever in an escapement we find the drops unequal we will likewise find that the shakes are unequal. A little experimenting will prove that it is quite possible to find drop present and the corresponding shake absent. Inequalities in drops or shakes will not ordinarily prevent a watch running, but a lack of either will cause stoppage. The manner of determining the amount of shake is explained in our tests.

131. *The Club Tooth Escape Wheel.*—When discusisng the pallets we learned that we had two types to consider, namely, the circular pallet, as used in American-made watches, and the equidistant, which is used only in the higher grades of imported watches. This distinction does not apply to escape wheels, as we have but *one* type of escape wheel readily adapted to either class of pallet.

132. *Escape Wheel Specifications.*—Number of teeth, 15; pallets to span 2½ tooth spaces. This from lock to lock equals 60 degrees; distance from heel of one tooth to heel of following tooth to be 24 degrees ($360° \div 15 = 24°$). To obtain the number of degrees suitable for combined width of tooth, pallet and angle of drop we divide 24 degrees by 2, the result being 12 degrees. Of this sum we assign 6 degrees for width of pallet, the remaining 6 degrees being divided as follows: For the width of tooth, 4½ degrees, and for the angle of drop, 1½ degrees. The draft angle or slant of the teeth is to equal 24 degrees, the lift or impulse angle to be 3 degrees.

133. *How to Draft an Escape Wheel.*—Let the line X B (Fig. 10) be the line of centers. With B as the center of the escape wheel and N as the radius, describe the arc C N C. This is the primitive or first circle of the escape wheel. Upon this circle the locking corners of all the escape wheel teeth will rest. On each side of the line B X lay off two angles each containing 30 degrees. The angle D B X and X B E each contain 30 degrees. Where the lines B D and E B intersect the arc C C we have marked T and U. Through these points draw the lines G A and H A tangent

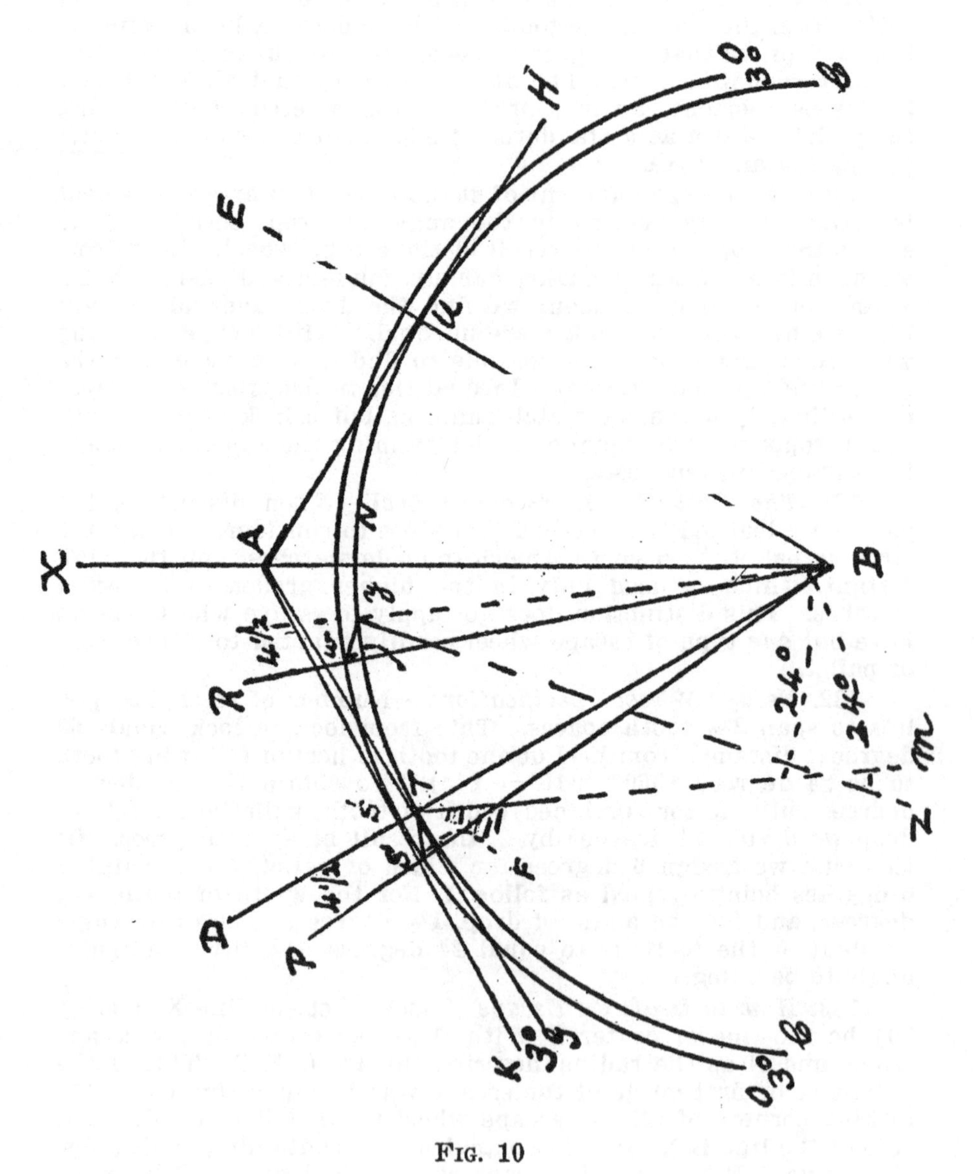

Fig. 10

to the arc C C . The tangent lines G A and H A meet on the line B X, namely, at A, this being the pallet center. Above the tangent line G A lay off an angle of 3 degrees as shown enclosed by the lines K A G. This is the angle of lift for the teeth of the escape wheel. Where K A intersects the line D B we have marked S′ With B as center and B S′ as radius draw the arc O O. This is known as the lifting or addenda circle. The space separating the arcs O O and C C will, when the teeth are correctly drawn in, provide for the lifting or impulse face. According to our specifications the lift on the teeth is 3 degrees. The toe of each tooth will rest on the arc C C, while the heel of each tooth will touch the arc O O. The width of a tooth is next drawn in. The required width being 4½ degrees, we lay out the angle P B D. Where P B intersects the arc O O is marked S; by drawing a line connecting S with T we define the tooth's lift or impulse plane. To mark out the slant of a tooth we place the center of the protractor at T, with the 90 degree mark extending along the line T B. We then count off 24 degrees as represented by the angle M T B. The line T M is the slant of the tooth, or, as previously mentioned, it may be called the draft angle of the tooth. The undercutting S to F is not obtained in obedience to any angle, the only rule which applies being that the undercutting should be of such extent that whenever the pallet dips into the wheel no contact of the parts is possible. To illustrate how other teeth can be drawn in we measure off 24 degrees on the arc C C away from the toe of the tooth already formed. From this point, marked Y, we draw the radial line Y B. From the point Y, away from the line Y B, an angle of 24 degrees is laid off, thereby defining the tooth's draft angle, as shown by B Y Z. Next an angle of 4½ degrees in width is laid out, its point of origin being the escape wheel center. The lines R B Y enclose the required 4½ degrees. Where the line R B cuts the arc O O we have marked W. Connecting the points W and Y gives us the lifting face of this tooth. If a student desires to make a drawing of an escape wheel showing the 15 teeth it is advisable first to space off the primitive circle into 15 divisions 24 degrees apart. Drawing in each tooth as spaced will yield an irregular drawing. Hence the advice: make the divisions first, then draw in the outlines of each tooth.

LESSON 7

THE LEVER—DRAFTING

134. *The Lever.*—The lever is a straight metal bar attached to or a part of the pallet arms. The end of the lever, known as the fork, is illustrated in Fig. 11.

135. *Form of Fork.*—The fork parts are the *horns*, the *corners* of the slot, or notch, and the *slot.* The horns (Fig. 11) are shown

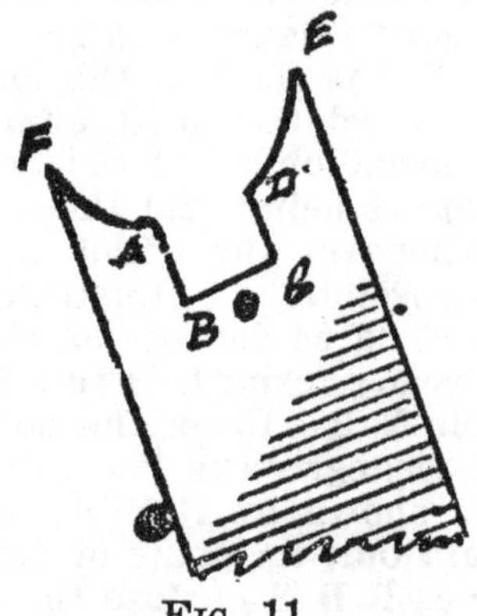

FIG. 11

from F to A and D to E. The corners of the notch are respectively marked A and D. The slot is enclosed by the lines A B, B C, C D.

136. *The Slot Corners.*—In a *single* roller escapement only *very short horns* on the fork are necessary. In fact, the main parts of the lever horns are the corners A and D, together with a very slight amount of horn to fully insure the soundness of the escapement action. The preservation of the safety action, by means of the slot corners, is intimately associated with the action of the roller jewel. This will be explained when we consider the question of the safety action.

137. *The Slot.*—The slot or notch is A B C D, illustrated in Fig. 11. Its purpose is that of receiving the roller jewel. The moment the roller jewel enters the slot it strikes one side of the notch a blow. The effect of this blow is, first, to lift the lever away from its bank; second, it causes unlocking of tooth and

pallet; third, just as unlocking takes place the lifting angles of tooth and pallet combined cause the opposite side of the slot to deliver a return blow to the roller jewel. This sets the balance vibrating with renewed energy. It is desirable to remember two points related to the above action; first, the speed of the roller jewel is checked and *decreased* by the force consumed in striking the unlocking blow; secondly, the speed of the lever through the energy developed by the lifts becomes *greater* than the speed of the roller jewel. In the lifts on tooth and pallet we have the reason why the opposite side of the slot delivers the return blow which keeps the balance vibrating.

138. *Angles of the Fork.*—The angles relating to the fork all originate from the pallet center. They are four in number, as follows: First, the angle governing the width of the slot; second, the angle of freedom for roller jewel when within the slot; third, the angle of freedom which controls the space separating the slot corners from the path of the roller jewel; fourth, we shall include with the fork angles that angle which separates the guard pin (lever against bank) from the edge of the table roller.

139. *Fork Specifications.*—Angular motion of lever, 10½ degrees; width of slot, 5 degrees. Acting length of lever to equal the distance between the centers of escape wheel and pallets. The angular motion of the lever consists of the following: Lift on tooth, 3 degrees; lift on pallet, 5½ degrees; total of drop lock and slide, 2 degrees.

140. *Drafting a Fork.*—Let A B (Fig. 12) be the line of

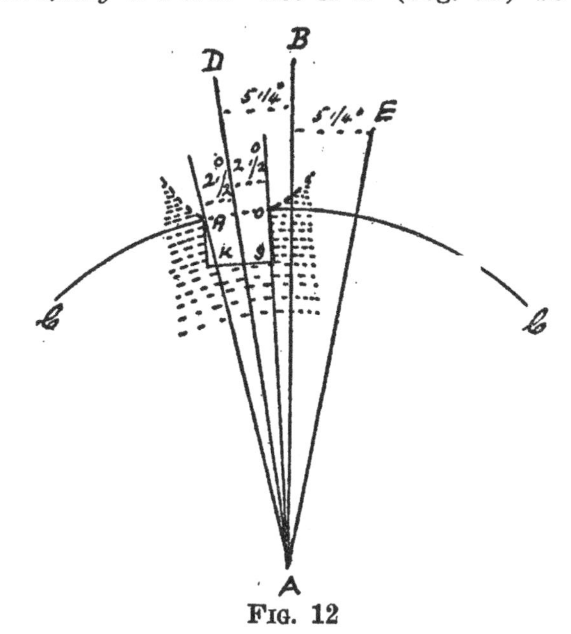

FIG. 12

centers, A being the pallet center. With A as center, and a radius equal to the distance between the escape wheel and pallets, draw the arc C C. Upon some part of the arc C C the slot corners will be located, consequently this arc defines the acting length of the lever. The angular motion of the lever equals 10½ degrees; therefore on each side of A B lay off two angles of 5¼ degrees each as shown by D A B and B A E. The width of the lever slot is given as 5 degrees. To define this angle lay off on each side of the line D A two angles, each possessing 2½ degrees. The whole angle H A O is therefore one of 5 degrees. H and O represent the slot corners, hence we draw in the lever slot as defined by H K S O. The horns are then drawn in to suit the requirements explained in the chapter on the safety action. The angle which provides freedom between the roller jewel and slot corners and the angle allowing freedom between the guard pin and the edge of the roller are both closely associated with the maintenance of the Safety action. These we shall consider in due course. (See Lesson 15.) In order to make certain points more clear and plain the drawings are illustrative in character. If drawn to scale their diminutive size would rather confuse than otherwise.

LESSON 8

THE ROLLER JEWEL

141. *The Roller Jewel.*—The roller jewel or impulse pin is a cylindrical-shaped jewel inserted into the table roller. About one-third of a roller jewel's cylindrical face is flattened off. The old style round jewel pin necessitated the opening of the bankings to an unnecessary extent. The result obtained by flattening the roller jewel is, the angular motion is lessened, or, to express it another way, the lever's motion from bank to bank is decreased.

142. *Action of a Roller Jewel.*—When a watch is running the roller jewel enters the lever-slot and strikes one side of the notch a blow. The force of this blow lifts the lever off its bank and unlocks the tooth and pallet; the escape-wheel tooth then enters on to the pallet jewel's impulse plane. The direct effect derived from the contact of the two lifting planes is to cause the opposite side of the slot to deliver a return blow to the roller jewel. It is this blow which causes the vibration of the balance. This action of giving and receiving a blow is kept up until the watch runs

143. *Angles Relating to the Roller Jewel.*—The width of the roller jewel is obtained by an angle whose starting point is the pallet center. The width of the roller jewel is naturally related to the width of the fork slot. The slot's width must always exceed that of the roller jewel, the difference between the widths being known as "the freedom of the roller jewel within the slot." The angle providing freedom between face of roller jewel and slot corners arises at the pallet center and is illustrated in Fig. 13 (A B C). This angle is of *great importance* in relation to the safety action, as we shall later explain. The roller jewel's angle of contact or arc of contact with the fork is technically spoken of as the impulse angle of the roller jewel. The distance from the balance center to the face of the roller jewel is termed the impulse radius or the roller-jewel radius. The distance between the balance and pallet centers is controlled combinedly by the angle relating to the lever's angular motion and the angle of impulse. Given these angles we can calculate the distance the pallet and balance centers should be apart. It has not been thought necessary to make a theoretical draft of the roller jewel's

position as it relates to the slot corners. The graphic drawing (Fig. 13) conveys all the practical information that one drawing can convey, which is, that a roller jewel requires a certain amount

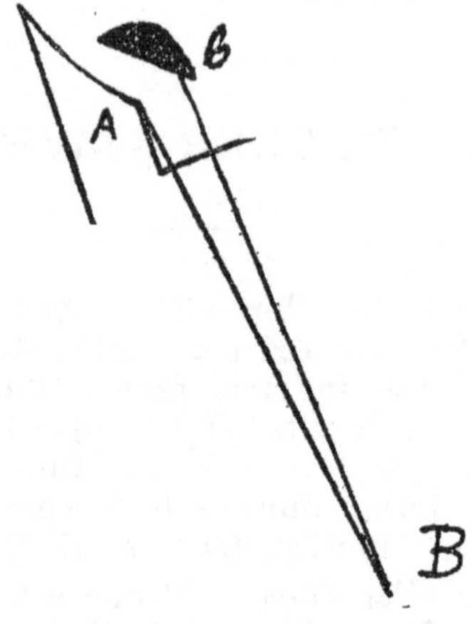

Fig. 13

of freedom when passing the slot corners under normal escapement conditions—by which we mean that slide is present. Our lessons, practical and theoretical, on banked to drop positions and on the safety actions will be found much more beneficial than any extended explanatory instruction on drafting a roller jewel in position. The same statement applies to instructions for the drafting of a table roller.

144. *The Table Roller.*—The table roller in an escapement possessing but one roller has two functions—first, to hold the roller jewel; second, its edge is an important feature of the safety action. In a double-roller escapement we have two rollers. The larger roller is termed the impulse roller and carries the roller jewel. The smaller table is known as the safety roller and is entirely associated with the safety action. The action of these rollers will be found explained in our chapters on the safety action.

145. *Roller's Angle of Freedom.*—The angle which governs the amount of freedom between the edge of the roller and the guard point originates at the pallet center. This angle is illustrated in Fig. 14 by the lines S A N; the amount or extent of this angle varies with the type of escapement. For a thorough understanding of the variations of this angle, either under normal or banked to drop conditions, the reader is referred to that portion of this book treating on the safety action and its tests. This also applies to the requirements of the various escapement types when banked

to drop, as hereafter described. The angle of freedom of the guard point from the edge of the roller is a matter of vital importance to the safe action of an escapement. It is also a

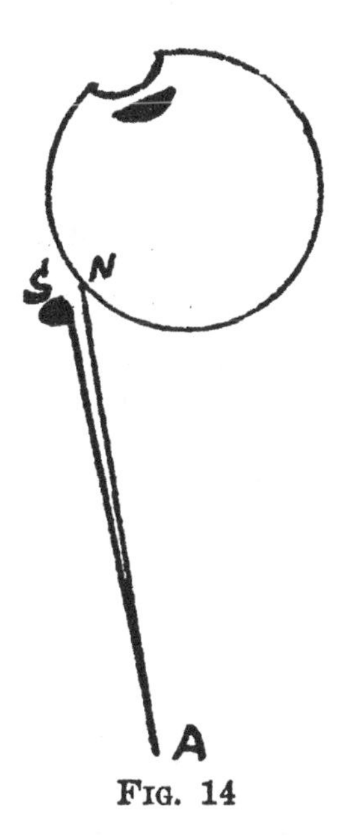

Fig. 14

lengthy and somewhat intricate subject, best mastered from practical experience as outlined in our tests.

146. *Crescent or Passing Hollow.*—The provision of a crescent or passing hollow cut into the edge of the roller enables the guard pin to pass from bank to bank without touching any part of the circumference of the roller. Its purpose is to insure the guard pin a free passage under normal conditions. Width and depth are its important features (see 198).

LESSON 9

DRAW AND ITS EFFECTS IN SINGLE AND DOUBLE ROLLER ESCAPEMENTS

147. *Draw.*—Draw is a result obtained from two sources—first, from the inclination or slant on the locking face of a pallet jewel; secondly, from the inclination of the tooth. Only the corner of the tooth should touch the locking face of the pallet. We might define draw as the force, or as the mechanical suction which under normal escapement conditions holds the lever against its bank. The cause of draw is located in the pallet and tooth action, aided by the power of the mainspring. The effect of draw is shown by the lever and its parts.

148. *Slide Lock, Its Relation to Draw.*—If we observe an escapement in action we will see the instant a tooth drops and locks on a pallet jewel that the pallet immediately starts to dip into the wheel, thereby increasing the amount of lock. This increase of the lock is spoken of as the slide or as the slide lock. This sliding lock of the pallet with the tooth is a product of the force termed draw. If the draw is imperfect—that is, weak and unable to satisfactorily retain the lever against its bank—the associated slide lock will be correspondingly ineffectual. The *extent* of slide is entirely controlled by the banking pins.

149. *Draw, Its Effects.*—The *amount* of draw necessary to hold a lever against its bank should be just sufficient to offset the ordinary body motions which a watch is subjected to in daily use. Should a watch receive an extra hard jolt, and the lever in consequence be thrown away from its bank and the guard pin comes in contact with the edge of the roller, the force of draw will promptly return the lever to its bank. Unnecessary friction between the guard pin and roller is thereby prevented. If the draw in an escapement is not strong enough to hold the lever against its bank, when subjected to the shocks received in daily usage, the lever in such an escapement will frequently be jarred away from its bank. Consequently an undersirable amount of contact of the guard point with the edge of the roller will result. This may cause stoppage, or at least the timekeeping qualities of the watch will be seriously impaired.

150. *Draw in Single-Roller Escapement.*—In a single-roller escapement the effect of draw is important in three positions.

(1) During such times when the guard pin is outside the crescent.

(2) When the guard finger is within the crescent.

(3) When the roller jewel is opposite the slot corners.

Regarding the *first*, should a watch receive a shock of sufficient violence to throw the lever away from its bank the guard pin will come in contact with the edge of the roller. The action, however, of draw immediately returns the lever to its bank, the result being that *steady* contact of the guard pin with the edge of the roller is prevented. With reference to our *second* statement, should the lever be thrown off its bank at the moment the guard pin enters the crescent a small portion of the curve of the horn will come in touch with the roller jewel. If the draw is effective the lever promptly returns to its bank. Our *third* item has reference to the possibility of the lever leaving its bank at the moment the roller jewel is passing the slot corner, in which event the roller jewel and slot corner come in contact. The action of draw should then pull the lever back to its bank. In this manner draw is a factor in the safety action.

151. *Draw in Double-Roller Escapement.*—In a double-roller escapement we have three different phases of escapement action wherein draw must be effective. These three positions exactly correspond with the requirements as before set forth for a single-roller escapement. They are:

(1) During the time that the guard finger remains outside the crescent.

(2) When the guard finger is within the crescent.

(3) When the roller jewel is opposite the slot corners.

As the above items are practically the same as already set forth in our foregoing statement on draw in a single-roller escapement, students will find no difficulty in understanding them, especially if they follow out in a watch the following experiments:

152. *Draw. Experiment No. 1.*—When the roller jewel is well *past* the end of the horn stop the watch by placing a finger on the balance; then with a fine tool, such as a watch oiler, lift the lever off its bank, thereby causing the guard pin to come in contact with edge of the roller. When the tool is removed, the draw, if sufficient, will pull the lever toward its bank.

153. *Draw. Experiment No. 2.*—At the moment the guard pin or finger arrives *just within* the crescent stop the watch and hold the parts in position. Next, lift the lever off its bank so as to produce contact of the lever horn with the roller jewel; this done, remove the tool. The action of draw should then be sufficient to pull the lever toward its bank.

154. *Draw. Experiment No. 3.*—Guide the roller jewel *opposite* the corners of the lever slot, then hold it there. Next, lift the lever off its bank, causing contact of the slot corners with the face of the roller jewel. Remove the tool from the side of the lever, and the draw, if good, will return the lever to its bank. If any defect in the draw is detected by the above experiments it should be further confirmed.

155. *Testing the Draw.*—To thoroughly examine the draw remove the balance. We should then expect to find, power being present, that the lever is at rest against its bank. To apply the test lift the lever slightly off its bank, then let it go. If the draw is right the lever promptly returns to its bank. Again lift the lever off its bank, but this time a little further than before. If the draw is sound the lever will again return to its bank. A third time lift the lever off its bank nearly to the point of unlocking, and, as before, when the lever is released it should return to its banking. Should the lever hesitate about returning to its bank, assuming that the watch is clean and freshly oiled and that the pivots of pallet staff and escape-wheel pinion correctly fit their respectice holes, then to overcome the want of draw the slant of the pallet jewel's locking face will have to be altered.

156. *Altering Draw.*—As a rule, want of draw is generally due to a pallet jewel being too straight. To overcome this want of slant the stone should be tilted in its setting; experiments are advisable. For instance, students should determine for themselves the effect of tilting a pallet jewel. If the pallet jewel experimented upon fits tightly in its seat substitute a thinner stone, or else cut out the walls of the seat; the original pallet jewel can then be tilted as desired. Changing the slant of a pallet stone necessitates investigating the *drops*, the *shakes* and the *locks* as directed in the following chapters. A great deal more could be written on the subject of draw, but given the hint that draw can be altered by changing the slant of the pallet jewel, students can by experimenting on old watches quickly master the correction of this defect.

Also remember when changing the slant of a pallet stone that the action of the "lifts" always demand attention. (See Lesson 11.)

LESSON 10

DROPS AND SHAKES

157. *Drop.*—Drop is the freedom allowed for the action of the pallet stones with the teeth of the escape wheel. When a tooth of the escape wheel is released from a pallet jewel there occurs a free motion of the wheel. This freedom of motion is termed "the drop." The drop or free flight of the wheel ceases the instant another tooth comes into contact with the locking face of the opposite pallet. In all lever escapements two kinds of drop are present, namely, outside and inside. These should be equal. As drop is a waste of energy and when excessive is injurious to close timing, it follows that a large amount of drop is not desirable. Theoretically 1½ degrees of drop is the standard, but in the majority of watches it exceeds the amount stated. In Lesson 13 students will find a table by means of which the amount of drop can be approximated.

158. *Shake.*—Shake, like drop, is divisible into two parts, viz., outside and inside shake. The "outside shake" should equal in amount the "inside shake." Any given shake, however, is always *less* than its related drop; for instance, the amount of inside shake is *less* than the amount of inside drop. Shake is less than drop because of the draft angle of the pallet. If a student will observe a tooth in the act of unlocking from a pallet jewel's locking face, a pushing back or recoil of the escape wheel will be seen, caused as just mentioned by the draft angle on the pallet stone. This recoil of the wheel lessens the freedom between the teeth and pallet jewels. Hence our statement "*shake is always less than drop.*" Therefore when alterations affecting the drops are made, the shakes must be given the consideration they require. It is quite possible for an escapement to possess drop, and shake be lacking or nearly so. It is also quite certain that a shortage of shake in an escapement will cause stoppage.

159. *Drop and Shake—Escape Wheel Defects.*—In high-grade watches possessing steel escape wheels we generally find the drops and shakes approaching perfection. It is mostly among the cheaper grades of watches, especially such as have brass escape wheels, that we find defects in either drop or shake, or both. As brass is

not rigid like steel, this is one cause of the trouble. The teeth in brass escape wheels will get out of shape; some are longer than others. The teeth also may not be at an equal distance apart, all of which complicates the watchmaker's problem of securing safe drop and shake.

160. *Drop and Shake—Drop Lock Defective.*—The first thing that demands attention when irregularities in the drops or shakes are discovered is to examine the locks. By lock we mean drop lock exclusively. If the locks are unequal they should be equalized. Whenever the drops and shakes are found unequal usually the locks are unequal. It therefore follows that the correction of irregular lock overcomes to some extent irregularities of both the drops and the shakes.

161. *Testing Outside Drop.*—To test the outside drop allow a tooth to be discharged *from the exit pallet.* The escape wheel then moves free. This free motion of the wheel is its outside drop. Outside drop ceases the moment another tooth is caught on the opposing locking face of the opposite pallet. (See paragraph No. 19.)

162. *Testing Outside Shake.*—The outside shake is always less than its corresponding outside drop, as explained elsewhere. To test the outside shake bring the tooth found at rest on the locking face of the *receiving pallet* down to that pallet's lowest locking corner, as represented by Fig. 19. Hold the parts in this position while you note the amount of *space* which separates the *back of the exit pallet* from *the point of the tooth just behind it.* The space seen represents the least freedom of the escape-wheel teeth outside the pallet jewels. (See paragraph No. 22.)

163. *Testing Inside Drop.*—To test the inside drop allow a tooth to become discharged from the *receiving pallet.* When this happens the escape wheel moves free of all contact. The free flight of the escape wheel is known as its inside drop. Inside drop ceases the instant a tooth comes in contact with the opposing face of the exit pallet. (See paragraph No. 18.)

164. *Testing Inside Shake.*—To learn the extent of inside shake bring the tooth found at rest on the exit pallet jewel's locking face down to the lowest locking corner of the pallet, after the manner shown in Fig. 20. Retain the parts in this position and observe the space separating the back of the receiving pallet from the *point* of the *tooth* just behind it. The space so seen represents the inside shake. The inside shake is always less than its corresponding inside drop. (See paragraph No. 21.)

165. *Correcting Drop and Shake When Tight Inside.*—To correct drop and shake when deficient inside we should try spreading the end of the pallet stones apart. At times it is best to spread or tilt both pallet jewels apart. More commonly the defective inside shake is cured by the tilting of one stone only.

When this is the case the question which pallet stone we shall tilt comes before us. The answer *in a practical way* is decided by examining the draw. A test of the draw on each pallet usually shows draw as less effective on one stone than on the other. Therefore the stone to be altered, when possible, is the stone showing the poorest draw. If the draw is sound on the exit pallet and deficient on the receiving, then in most instances by tilting the end of the receiving stone away from the opposite pallet the draw can be increased. We must also remember that titling the pallet jewel affects the lock, the drop, the shake and the lifts. A little experimenting will prove these statements. (Consult paragraphs Nos. 437 and 438.)

166. *Correcting Drop and Shake When Tight Outside.*—If the drop and shake are tight or deficient outside we, as before, first see to the drop locks. Should it be desirable to use other means than attempting to make a correction by directly altering the locks we can do so by bringing the pallet jewels closer together, after the manner already described for correcting shake and drop when too tight inside. (See paragraphs Nos. 165-167.)

167. *Providing a Safe Amount of Shake by Altering a Pallet Jewel.*—To provide a safe amount of shake it becomes necessary at times to replace a thick pallet jewel with a thin one. When the watch is of a poor type the same effect, namely, substituting a thin for a thick stone, can be obtained by means of a diamond lap. By using such a tool a part of the back portion of a pallet jewel can successfully be ground away. This thins the stone at the required place and provides the requisite shake. The foregoing in a general way outlines the procedure to be followed. The points to be aimed for when tilting, shifting or changing the thickness of a pallet stone are to equalize the drops, the shakes, the locks and the draw. (Also, consult paragraph No. 171.)

LESSON 11

LIFT ON TOOTH AND PALLET, CORRECT AND INCORRECT

168. *The Lifts.*—The question of lift on tooth and pallet is an entirely practical one. An irregular action of the lifts prevents timing and causes stoppage. The student is therefore advised to make a study of the lifts at different phases of the escapement action. The manner in which a tooth first enters on to a pallet's lifting plane, their central relationship and their relative positions when disengaging should be closely studied and understood.

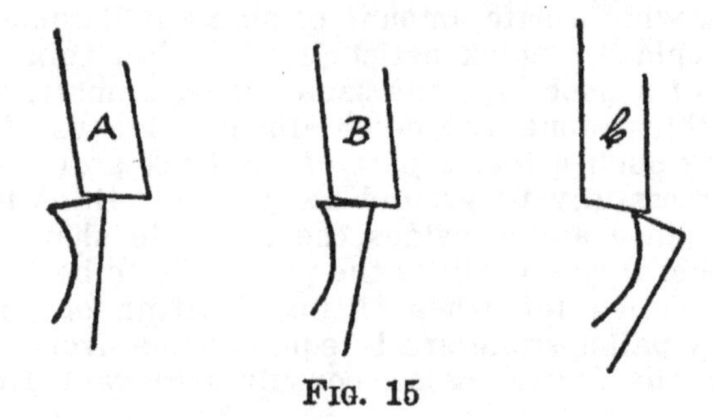

Fig. 15

169. *Correct Lift.*—The drawings (Fig. 15) A, B and C indicate the average relative positions of a tooth as it passes over the impulse face of the receiving pallet. The drawings (Fig. 16)

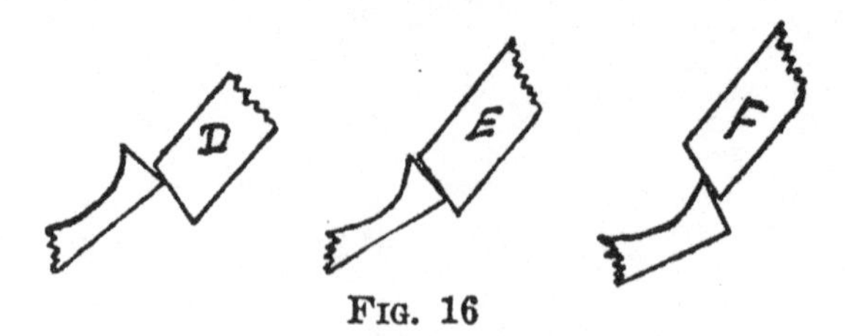

Fig. 16

D, E and F show the average relative positions of a tooth as it travels over the discharging pallet's impulse face.

170. *Incorrect Lift.*—Not infrequently we encounter lifting actions of an irregular nature, such, for instance, as illustrated in Fig. 17. This drawing shows that the pallet acts on the tooth

Fig. 17

in place of the tooth acting on the pallet. In Fig. 18 we have represented a disengaging action wherein we again find that the pallet corner is scraping the tooth's lifting plane. Errors such as

Fig. 18

shown in Figs. 17 and 18 must never be allowed to go uncorrected. Watches possessing this fault are a source of worry to untrained watchmakers and are entirely unsatisfactory to their owners.

171. *Correcting Lifting Errors.*—Errors in lift are generally attributable in old watches to mismatched parts. When the trouble is due to an unsuitable pallet jewel it should be replaced by one of correct form and make. Irregularities of the lifts are sometimes discovered in new watches. When such is the case the watch should be returned to the factory for correction. Our own experience is that the factory simply cuts the sides of the container holding the pallet jewel; this allows the stone to be so tilted that the error in lift is overcome. As watchmakers are not lapidaries, cutting the seat is their only solution of the problem. *Changing the slant* of a pallet jewel always brings with it questions of draw, drop, shake and lock. With these subjects we must be thoroughly familiar, including the topic of "Lift."

LESSON 12

TOTAL LOCK—DROP LOCK—SLIDE LOCK—BANKED TO DROP—BANKING PINS

172. *The Locks.*—It is *very important* that students verify in an escapement the statements in this and other paragraphs presented in Lesson 12. They contain practical essentials that

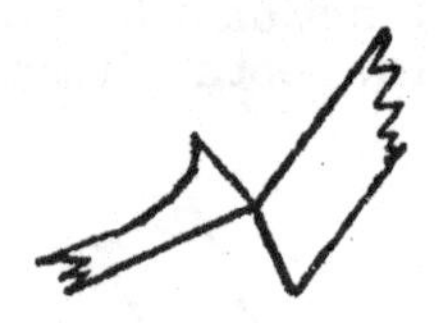

FIG. 19

students must be familiar with. The *total lock* of a tooth of the escape wheel on a pallet jewel is composed of two items, viz., *drop lock* and *slide lock.*

173. *The Drop Lock.*—Drop lock takes effect the instant a tooth drops on to the locking face of the pallet jewel. The extent or amount of drop lock is estimated from the lowest locking

FIG. 20

corner of the pallet up to that point on the pallet's locking face where the tooth dropped. Let B (Fig. 21) represent the point upon which the tooth dropped, then the distance from B to A shows the extent of drop lock. Drop lock is not a product of the banking pins. The banking pins simply mark out, when so adjusted, the position where drop lock takes place. The extent of drop lock is entirely due to the position in which we place the pallet jewels—that is, *in* or *out*, as the condition of the escapement may require. Book knowledge of drop lock, especially of the theoretical variety, is of little benefit to the student, because drop

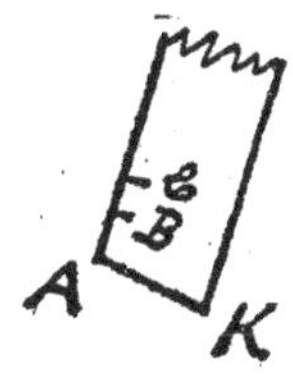

Fig. 21

lock is not a separate entity. It is one part of a whole, one link in the escapement. It is a varying quantity in nearly every watch, as every experienced watchmaker realizes. Drop lock, light or deep, is a question reaching further than the mere fact of lock itself. The tests and lessons to follow will bring this out clearly. For practical reasons we shall ask students to work out drop lock problems in connection with our tests. Thereby they can gain a practical knowledge of "correct lock."

174. *Slide Lock.*—Slide lock is the after or secondary lock which, when the bankings are opened out, follows drop lock. In Fig. 21 B to C represents the slide lock. Slide lock is therefore an after effect following drop. Slide lock and draw combinedly cause the pallet to dip deeper into the wheel. The extent of slide is entirely controlled by the banking pins. Every escapement in normal running condition must show slide. Therefore after making an escapement test under banked-to-drop rules it is necessary to restore the slide. This is done by spreading the bankings. Why slide is necessary will be better understood when the tests and safety actions are studied. We shall make this brief statement regarding it: By providing an escapement with slide it insures *greater* separation of the guard point from the edge of the roller when the lever rests against its bank.

175. *Total Lock.*—The *total* lock of a tooth on a pallet jewel is the sum of the drop lock plus the slide lock. In Fig. 21 A to B is the drop lock and B to C the slide; therefore A to C represents the total lock.

176. *Remaining or Safety Lock.*—This is best learned by making the following experiment: Bring the guard point of a sound escapement into contact with the edge of the roller, then observe the tooth and pallet. Whatever amount the tooth remains locked on the pallet jewel's locking face represents the remaining

or safety lock (Fig. 22). If the safety lock is wanting, that is, if the tooth enters on to the pallet jewel's impulse face (Fig. 23) while the guard point is *held in contact* with the edge of the

FIG. 22

table roller, then the error known as tripping is present. This defect we shall treat on in due course.

177. *The Three Safety Locks.*—In every escapement a safety lock is required as follows: (1) When the guard point touches

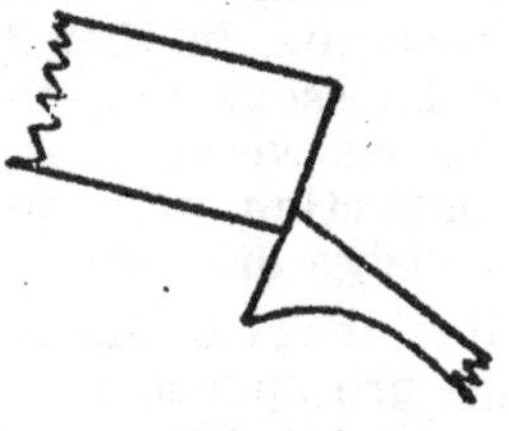

FIG. 23

the roller. (2) When the roller jewel touches the curve of the lever horn. (3) When the slot corners of the lever and roller jewel are brought into contact. For further particulars see guard, corner and curve safety tests.

178. *The Banking Pins.*—The banking pins are the two eccentric pins placed on each side of the lever. By means of the banking pins we can increase or decrease the slide lock. To a certain extent we can adjust them to control the distance between the guard point and the edge of the roller. With them we can also, within certain limits, control the extent of the roller jewel's contact with the walls of the lever slot. We can also adjust them to mark out that important position termed "banked to drop."

179. *Banked to Drop.*—The expression "banked to drop" conveys the fact that both banking pins are so adjusted that immediately a tooth drops and locks on a pallet jewel the lever that instant meets its bank. In an escapement of the Elgin type, when banked to drop, *a slight space* or freedom will be found separating the guard point from the roller. In escapements of the South Bend type, when banked to drop, the guard point will be found

in contact with the edge of the roller. All escapements *perfectly* banked to drop show *no* slide and, consequently, *no* run of the lever. This condition, especially in cheaply made watches, is unattainable because of the varying lengths of the teeth of the escape wheel. Draw, however, is and should be present. When we find an escape wheel with teeth of irregular length perfect banked-to-drop conditions are *impossible* to obtain. Some of the teeth will be exactly banked to drop; others will show slide. The defect cannot be overcome without a change of escape wheels. This rarely pays where cheap watches are concerned.

Banked to drop is the *key* which unlocks escapement difficulties. It will be found a splendid help in unraveling escapement errors.

LESSON 13

MEASUREMENT OF LOCK—APPROXIMATING—DEGREES OF LOCK AND DROP

180. *Measurement of Lock.*—The measure of one degree of of lock depends entirely on the size of the circle of which the degree is the 1-360th part. To learn how much one degree measures we must be acquainted with the length of the radius. In other words, we must know the measure of the distance separating the lowest locking corner of the pallet jewel from the center of the pallet staff. Given the measure of this distance we can, by means of the following rule, calculate the size of one degree of lock:

Example—The distance from the lowest locking corner of the pallet jewel to the pallet staff center measures 2.5 millimeters. Find the measure of one degree of lock.

Rule—Radius $\times$ 2 $\times$ 3.1416 $\div$ 360 = measure of 1° of lock.

$$2.5 \times 2. = 5.$$
$$5 \times 3.1416. = 15.7080.$$
$$15.7080 \div 360. — .044.$$

The answer—.044 millimeters is the measure of one degree of lock in this escapement.

181. *Approximating Degrees of Lock in Any Escapement.*—The watchmaker should cultivate his eye to estimate the number of degrees of lock of a tooth on a pallet. As an aid toward this we divide up the width of a pallet jewel. Assuming that the full *width* of a pallet (see Fig. 21, A to K) approximates 10 degrees of *lock*, then if a tooth is *locked* on a pallet to an extent equalling one-half the *width* of the pallet we are safe in saying that the said tooth is locked on the pallet jewel to the extent of five degrees. The width of a pallet means the distance across the stone's impulse face from its entrance to its exit corner. Students must not confuse an estimation of the extent of the *drop* with an estimation of the amount of *lock*. As regards the lock, one-fifth of the pallet's width equals two degrees of lock, but when we estimate *the drop*, then one-fifth of the width of the pallet jewel equals *one* degree of drop. (See paragraphs Nos. 183 and 184.)

182. *Estimating the Angle of Drop.*—We can estimate the angle of drop in an escapement as before, by considering the *width* of the pallet as our standard of measure. The angle of drop and the angle of controlling the width of a pallet jewel are measured from the *same* point, namely, the escape wheel center; this makes them closely related. Therefore accepting the width of a pallet as five degrees, we can readily realize that if a tooth drops to an extent equalling one-half the width of the pallet the drop will equal 2½ degrees. In this maner it is easy to approximately estimate the angle of drop in an escapement as well as the amount of shake.

183. *Table for Approximating Degrees of Lock.*

1/10 width of pallet equals 1° of lock.
1/5 width of pallet equals 2° of lock.
¼ width of pallet equals 2½° of lock.
½ width of pallet equals 5° of lock.
¾ width of pallet equals 7½° of lock.
1 width of pallet equals 10° of lock.

184. *Table for Approximating Degrees of Drop.*—

1/10 width of pallet equals ½° of drop.
1/5 width of pallet equals 1° of drop.
¼ width of pallet equals 1¼° of drop.
½ width of pallet equals 2½° of drop.
¾ width of pallet equals 3¾° of drop.
1 width of pallet equals 5° of drop.

The basis of the above figures accepts a pallet width of 5 degrees as its standard. In an earlier lesson on pallet drafting the pallet width was given as 6 degrees. As a matter of fact, the width of pallet generally used with escape wheels having club teeth varies from 5 to 6 degrees. As an estimate of the lock or drop made simply by an observation can only be approximate, the figures in the above table are therefore sufficiently close to apply to either width of pallet.

184 A. *Approximating Degrees of Shake.*—As shake and drop are intimately related, we can, when necessary, apply the above table of drop to estimate the degrees of shake.

LESSON 14

ROUTINE ACTION OF THE SINGLE ROLLER ESCAPEMENT—IMPORTANT GUARD PIN POSITIONS

185. *Routine of Escapement Action.*—If we observe an escapement in action under normal conditions, namely, with slide present, and we commence our investigations at the time the lever is at rest against its bank, a tooth locked on pallet jewel and the roller jewel starting on its return journey toward the slot, an analysis of the routine action of the escapement would read as follows:

First Observation—Tooth locked on the pallet.

Second Observation—The roller jewel is traveling toward the lever slot.

Third Observation—Just before the roller jewel *starts to enter the slot* the guard pin enters the crescent.

Fourth Observation—The roller jewel enters the slot and strikes one side of the slot a blow.

Fifth Observation—The blow delivered by the roller jewel against the side of the lever slot causes unlocking of tooth and pallet, thereby allowing the unlocked tooth to slip on to the pallet jewel's impulse face.

Sixth Observation—Through the meeting of the lifting planes, of tooth and pallet, aided by the power of the mainspring, the opposite side of the lever slot delivers a blow to the roller jewel.

Seventh Observation—This blow causes the balance to rotate with renewed energy.

Eighth Observation—When the tooth left the discharging corner of the pallet it dropped (drop) and another tooth locked on the opposing pallet (drop lock).

Ninth Observation—When drop lock takes effect the guard pin theoretically is within the crescent, but practically considered, owing to the velocity of the parts, the guard pin is leaving the crescent. It is at this phase of the escapement action that the guard pin comes *closest* to the edge of the roller. The position of the guard pin just mentioned must always be kept in mind by the repairer, especially when circumstances require us to adjust the guard pin a trifle closer to the edge of the roller than is really desirable.

Tenth Observation—Immediately drop lock takes place the pallet slides deeper into the wheel (slide lock). The deeper the pallet slides into the wheel the further the guard pin is carried away from edge of the roller.

Eleventh Observation—Coincident with the increase of the slide lock in the run of the lever toward its bank. When the lever arrives at its bank the guard pin is then furthest from the edge of the roller. The cause of the effects noted above, viz., slide and run, is entirely attributable to draw.

Twelfth Observation—The amount of slide lock, the amount of run and the distance separating the guard pin from the edge of the roller are three correlated effects. The amount of each is normally controlled by the position of the banking pins.

186. *Important Guard Pin Positions.*—We wish to impress on students the following three positions of the guard pin. It is advisable that actual observation of the described actions be followed out in an escapement:

First Position—When the roller jewel enters the notch, preparatory to striking the unlocking blow, the guard pin is within the crescent.

Second Position—As a tooth of the escape wheel drops on the locking face of the pallet jewel the guard pin, for all practical purposes, owing to the velocity of the parts is barely outside the crescent. This means that the guard pin is only clear of the roller edge and no more. As previously mentioned, this represents the closest the guard pin aproaches the roller edge when an escapement is in action.

Third Position—The next position of the guard pin is when the lever is at rest against its bank. Slide lock being present, the guard pin is then at its greatest distance from the edge of the roller.

LESSON 15

THE SAFETY ACTIONS OF THE SINGLE-ROLLER ESCAPEMENT—GUARD PIN—OVERBANKING—TRIPPING

187. *The Safety Action.*—The purpose of the safety devices is to insure the escapement continuing in action should the watch receive a shock of sufficient force to throw the lever off its bank, in which event the parts relating to the safety action come in contact and thereby assist in returning the lever to its bank. In a single-roller escapement the parts composing the safety action are the guard pin, coassociated with the edge of the roller; the roller jewel, coassociated with a very short part of the lever horn, and the roller jewel as associated with the corners of the slot. Most intimately related to the above combinations are the lock of the escape wheel tooth on the pallet jewel's locking face and the draw.

188. *The Guard Pin's Safety Actions.*—The office of the guard pin as a safety action factor is as follows:

(a) To prevent overbanking.

(b) To prevent tripping.

(c) To prevent contact of the roller jewel with the greater part of the lever horn. This it does up to the moment the guard pin enters the crescent.

Once the guard pin enters the crescent the preservation of the safety action then depends upon the roller jewel, associated with either a small part of the horn or the slot corner, as described in the following:

189. *The Roller Jewel's Safety Actions.*—The purpose of the roller jewel as a factor in the safety action of single-roller escapements is confined to its association with the slot corners and that part of the horns close to the slot corners. Its function as *a part* of the safety action is to prevent tripping.

190. *Overbanking.*—As previously stated, one of the functions of the guard pin is to prevent overbanking. The shocks and jolts a watch receives in the course of every-day usage will at times, when sufficiently violent, jar the lever away from its bank and result in contact of the guard pin with the edge of the roller. If

conditions are correct the lever returns to its bank. This, of course, relieves the pressure of the guard pin against the roller. If conditions are incorrect the guard pin slips past the unbroken edge of the table. In this event the lever passes to its opposite bank and the roller jewel on its return excursion, instead of being able to enter the fork, strikes on the outside of the horn. The escapement is then in the condition usually termed "overbanked." In the ordinary course of an escapement's action the lever should never move from one bank to another except when under the control of the action of the roller jewel in the lever slot. Should the lever jump from one bank to another in an irregular manner the escapement is put out of action and overbanking results. Overbanking, therefore, implies an irregular motion of the lever from bank to bank without the aid of the roller jewel. It also implies that the safety action failed in its function. Failure of the safety action sufficient to allow overbanking to take place is attributable to one or more causes.

191. *Causes of Overbanking.*—Overbanking is due to some of the following defects: A guard pin too far away from the edge of the table. A loose guard pin. The edge of the roller running out of truth; sometimes this is the result of an attempt to close the hole in the roller in an effort to securely fasten it on the balance staff. Bent pivots of the balance staff also produce an eccentric motion of the roller. A frequent cause of overbanking is that of holes too large for the pivots of the pallet or balance staff. Jewels loose in their settings, or settings loose in their seats. Defective draw, with uncertain adjustment of some parts of the safety action, will at times be responsible for an overbanking error.

192. *Tripping.*—Tripping is the act of a tooth of the escape wheel leaving the locking face of the pallet jewel in an irregular manner. Any unlocking of tooth and pallet not caused by the action of the roler jewel with the slot is an irregular unlocking. All watches should be subjected to tests for tripping errors and corrections made if an error is found. We have *three* positions in a single-roller escapement wherein tripping errors may develop, as follows:

(a) While the guard pin is outside the crescent.
(b) When the guard pin just enters the crescent.
(c) When the roller jewel is opposite the slot corners.

TRIPPING TESTS (SINGLE ROLLER)

193. *Guard Safety Test.*—First—Place a finger on the balance rim and rotate the balance so as to bring the roller jewel beyond the end of the horn.

Second—Hold the balance steady, then with a fine broach lift the lever off its bank, thereby bringing the guard pin and roller in contact.

Third—Retain the parts in position and with an eye glass note the amount of the remaining or safety lock of the tooth on the pallet.

Fourth—If the tooth remains on the locking face of the pallet jewel (Fig. 22) the safety action, so far as tested, is sound. If a tooth enters on to the pallet jewel's impulse face (Fig. 23, irregular unlocking) the error known as tripping is present and calls for correction. The cause of the error might be attributable to the drop locks being too light or the guard pin not being correctly adjusted to the roller. With the assistance of the tests to follow students will be able to locate the cause of error. Should we discover at any time a very light safety lock, make a test of all the teeth, for the reason that some of the escape wheel teeth may be shorter than others. If so, look out for tripping errors. Slight trips cause irregular stoppage of the escapement, hence we repeat our advice, when the safety lock is extremely light test all teeth of the escape wheel.

194. *Corner Safety Test.*—First—Rotate the balance so as to bring the roller jewel opposite one of the slot corners.

Second—With a fine tool lift the lever away from its bank sufficiently to cause contact of the slot corner with the roller jewel.

Third—We note the condition of the remaining or safety lock. The tooth must be found on the locking face of the pallet jewel. If a trip is discovered, it must be corrected. The cause might be due to the lock, the position of the roller jewel, or to the acting length of the lever. The nature of the error should be uncovered by means of the angular and corner tests, as explained elsewhere in the lessons.

195. *Curve Safety Test.*—First—Place a finger on the balance and guide it so the guard pin just enters the crescent.

Second—Hold the parts in this position and with a broach or watch oiler lift the lever off its bank, thus causing contact of the roller jewel with a small part of the horn near the slot corners. How much of the horn can thus be found in contact with the roller jewel *depends* on the size of the crescent.

Third—Still hold the parts in contact while with an eye glass the remaining or safety lock of tooth on pallet is inspected.

If the relationship of the parts is correct a tripping error will not be discovered. As a matter of fact, in single-roller escapements this part of the test can be omitted, because tripping errors rarely develop here. The positions we must test are the guard pin against the roller and the roller jewel with the slot corner, as stated below.

LEVER HORNS AND CURVE TEST (Single Roller)

196. *Relation of Horns to Roller Jewel.*—In connection with our subject a brief repetition of the relation of the lever horns to

the roller jewel is desirable. The most perfect relationship when subjected to test conditions is that of the non-contact of the roller jewel with the lever horns while the guard pin *remains outside* the crescent. Once the guard pin enters within the crescent and we apply tests, contact of the roller jewel with a very short portion of the horn is to be expected. This is one of the features of the safety action. Again, when the roller jewel under test conditions is brought in contact with the slot corners no decided catch of the parts should be found. The roller jewel under test conditions should rub evenly along the short part of the horn and past the slot corner without showing any inclination to stick. Undue friction developing into a catch of the parts must be overcome. The following tests show the relationship of the curve of the horn to the roller jewel.

197. *Curve Tests.*—First—Place a finger on the balance and guide the roller jewel into a position beyond the tip of the lever horn.

Second—With a tool lift the lever away from its bank and maintain contact of the guard pin with the edge of the roller.

Third—With all parts held as directed, slowly commence rotating the balance, thereby bringing the roller jewel past the horn. No contact should be felt. If contact *is* detected it should be only of the slightest character while the guard pin is *outside* the crescent.

Fourth—Once the guard pin enters the crescent a slight rub will be felt of the roller jewel on the horn and on the slot corner as it passes. No catching of the parts is permissible, for such a defect could produce stoppage of the watch.

These tests will be found in more concise form in Lesson 31.

LESSON 16

ROUTINE ACTION OF THE DOUBLE-ROLLER ESCAPEMENT

198. *Routine Action of the Double-Roller Escapement.*—If we observe the routine action of a double-roller escapement when running under normal conditions, namely, with slide present, the following would be a statement of the actions. Assuming that the roller jewel is beyond the horn and the lever at rest against its bank:

First Observation—Tooth locked on pallet.

Second Observation—When the guard finger enters the crescent the roller jewel is opposite the tip of the horn.

Third Observation—The guard finger is well within the crescent when the roller jewel enters the slot and strikes the unlocking blow.

Fourth Observation—The blow delivered by the roller jewel to the lever slot caused unlocking of tooth and pallet. In consequence the tooth entered on to the pallet jewel's impulse face.

Fifth Observation—The effect of unlocking resulted as follows: The lifting plane of the unlocked tooth, through power derived from the mainspring, pushed its way over the impulse face of the pallet jewel. The contact of the lifting planes caused the opposite side of the slot to deliver a return blow to the roller jewel.

Sixth Observation—The effect of the return blow causes the balance to vibrate with renewed energy.

Seventh Observation—The moment a tooth left the discharging corner of the pallet it dropped (drop) and another locked on the opposite pallet (drop lock).

Eighth Observation—When drop lock took effect the guard finger was deep within the crescent.

Ninth Observation—Immediately on completion of the drop lock the pallet commenced to dip into the wheel (slide lock). As a result the lever runs toward its bank.

Tenth Observation—The guard finger is far *within* the crescent when the lever starts to run toward its bank.

Eleventh Observation—When the guard finger emerges from

the crescent the distance separating the guard finger from the edge of the safety roller has been increased by the amount of slide lock. To state the foregoing another way, when the guard finger emerges from the crescent the lever is at rest against its bank and the roller jewel will be traveling toward the extremity of the horn.

NOTE.—The *width* of crescent and *length* of horn are closely related. For instance, when the guard finger *just* emerges from the crescent, the length of horn must be such, that its tip is opposite the center of the roller jewel. A horn of shorter length will cause trouble.

LESSON 17

THE SAFETY ACTION OF THE DOUBLE ROLLER ESCAPEMENT—GUARD FINGER—ROLLER JEWEL—OVERBANKING—TRIPPING

199. *Safety Action—Parts of the Double-Roller Escapement.*—In a double-roller escapement the parts comprising the safety action are:

(a) The lever horn coassociated with the roller jewel.

(b) The slot corners coassociated with the roller jewel.

(c) The guard finger coassociated with the edge of the safety roller.

Closely allied to the above is the lock of the tooth on the pallet jewel and the draw.

200. *The Guard Finger's Safety Actions.*—The function of the guard finger in a double-roller escapement is a preventive one, as follows:

(d) To prevent overbanking

(e) To prevent tripping.

(f) To prevent the roller jewel touching the tips of the horns.

201. *The Roller Jewel's and Guard* Points' *Safety Actions.* The office of the roller jewel as a factor in the escapement action of a double-roller escapement is given below:

(g) When the guard finger just enters the crescent should the lever from any cause be thrown off its bank the roller jewel will meet the face of the horn and thereby prevent tripping.

(h) When the roller jewel is opposite the slot corner should the lever at that moment be thrown off its bank the slot corner will come in contact with the face of the roller jewel. This prevents tripping and insures soundness of this part of the safety atclou.

It can be gathered from the above that just as long as the guard finger remains *outside* the crescent the protection of the safety action belongs to the guard finger and the edge of the safety roller. Once the guard finger *enters* the crescent it is of *no further use* as a factor in the safety action.

(k) When the guard finger of a double-roller escapement enters the crescent the preservation of the safety action is due to:

First—The curve of the horn meeting the roller jewel.

Second—The corner of the lever slot meeting the roller jewel.

202. *Overbanking.*—As the causes of overbanking in a double-roller escapement are similar to those already described in our treatment of this error in a single-roller escapement it will be unnecesasry to repeat it here.

203. *Tripping.*—In a double-roller escapement we have *three positions* wherein to suspect the existence of a tripping error:

(a) While the guard finger is opposite any part of the edge of the safety roller *outside* the crescent.

(b) When the guard finger *enters* the crescent and the roller jewel is opposite any of the central part of the horn

(c) When the roller jewel is opposite the slot corners.

The tests employed for proving or disproving the existence of a tripping error and for determining if length of horn is correct are stated in the following:

TRIPPING TESTS (Double Roller)

204. *Guard Safety Test.*—To discover if a trip is possible while the guard finger remains *outside* the crescent, commence by rotating the balance so as to place the roller jewel at some point beyond the end of the horn. This done, lift the lever off its bank, thus causing contact of the guard finger with the edge of the safety roller. Hold the parts in contact and use an eye-glass to observe the lock of the tooth on the pallet jewel. If the safety action is sound, the tooth will be found locked on the pallet jewel's locking face. If a trip is discovered, that is, if the tooth leaves the locking face and enters ever so slightly onto the stone's impulse face the cause must be determined and the error corrected.. If the error is due to a short guard finger, the finger may be stretched to correct the error. If the cause is due to defective drop lock, it is an easy matter to see if the lock is too light and to make the necessary alteration. Consult article on "Tripping in a Single Roller Escapement."

205. *Corner Safety Test.*—By bringing the slot corner in contact with the roller jewel and then examining the condition of the safety lock, it can be learned if the remaining lock is sound. If the escapement trips corrections are of course necessary.

206. *Curve Safety Test.*—When the guard finger is brought in contact with the edge of the safety roller and we rotate the balance, so as to cause the roller jewel to stand opposite the tip of the horn, no contact of the roller jewel with the end of the horn is permissible. The parts mentioned should be free from each other. If we continue rotating the balance, still maintaining the guard finger pressed against the side of the roller, we will

find that the moment the guard finger enters the crescent that the curve of the horn will come in direct contact with the face of the roller jewel. Having thus obtained contact of the roller jewel with the curve of the horn, take an eyeglass and observe the position of the tooth on the pallet jewel. If this part of the safety action is sound, the tooth will be found on the *locking face* of the ballet jewel. An *incorrect finding* would be to discover the tooth on the impulse face of the pallet stone. This means the escapement trips. Changes are then necessary or the watch will stop when in daily usage.

LEVER HORNS AND CURVE TEST (DOUBLE ROLLER)

207. *Testing the Length of the Horn.*—To learn if the horn of the lever in a double roller escapement is of correct length, lift the lever off its bank, causing the guard finger to come in contact with the edge of the safety roller. Keep the parts in contact and guide the balance so that the center of the roller jewel stands opposite the end of the horn. When the roller jewel stands in this position the *guard finger* will be *just outside* the crescent. Therefore, the length of the horn and the size of the crescent are directly related. If the width of the crescent is increased, the length of the horns must likewise be increased to meet the required conditions, viz., when the roller jewel stands centrally opposite the end of the horn, the guard finger must be just outside the crescent as before stated. When this specification has been met the lever horns are of correct length.

208. *Curve Test.*—*First*—Place a finger on the balance rim and guide the roller jewel into position beyond the end of the horn.

Second—With some fine tool lift the lever away from its bank, then hold the guard point in contact with the edge of table.

Third—With finger on balance rim, and guard point kept in contact with table's edge, slowly turn the balance, thereby bringing the roller jewel past the tip of the horn. No contact of the roller jewel with this part of the horn is permissible. Should any be detected and the parts stick or catch alterations are desirable, as instructed in the Test Lessons.

Fourth—At the moment the guard point enters the crescent the horn and roller jewel come into contact and remain so until the roller jewel enters the lever slot. Regarding this contact, the roller jewel should slide smoothly over the face of the horn and past the slot corner without showing any tendency of the parts to stick, otherwise stoppage of the watch may result.

LESSON 18

SOURCES OF ESCAPEMENT ANGLES

209. *Relationship of the Angles.*—All angles of freedom are measured from the pallet center, the angle of drop excepted. Hence with this exception the angles of freedom are correlated.

The angle of lock is likewise measured from the pallet center. This angle is therefore correlated to the angles of freedom which originate at the pallet center. Escapement angles having a *common source* are affected by the alteration of one of their number; for instance, if we alter the angle of lock, the angles pertaining to the safety action reflect the change. For this reason carefulness is counseled before vital alterations are made in an escapement.

210. *Source of Escapement Angles.*—Judging from the above remarks, it is advisable for students to learn the source of the various escapement angles, so that previous to an alteration being made, a definite opinion can be formed as to the effect of the proposed change on some correlated part.

211. *Angles Radiating Toward the Fork—Their Point of Origin, the Pallet Center.*—A—Angle of freedom of the roller jewel from the slot corners.

B—Angle of freedom of the guard-point from the edge of the roller.

C—In a double-roller escapement the angle of freedom regulating the distance between the curve of the horn and the path of the roller jewel.

D—The angle of freedom which provides space between the end of the horn and the path of the roller jewel.

E—The freedom angle of the roller jewel when the roller jewel is contained within the slot.

212. *Angles Radiating Toward the Tooth and Pallet—Their Source of Origin, The Pallet Center.*—F—The angle of lock of the tooth on pallet.

G—The angle of lift on the impulse face of the pallet jewel.

H—The angle of lift on the impulse plane of the tooth.

213. *Angles Arising at Escape-Wheel Center.*—K—The angle controlling the width of the pallet jewel.

L—The angle regulating the width of tooth of the escape wheel.

M—The angle controlling the amount of drop.

214. *Source of Other Important Angles.*—N—The angle of draft which determines the slant of the pallet jewel's locking face. It originates at the lowest locking corner of a pallet stone.

O. The angle which provides the slant found on the locking face of a tooth. The angle governing this slant starts from the tooth's locking corner. In earlier lessons all of the angles above mentioned can be traced in the various drawings and reading matter connected with same.

LESSON 19

THE SAFETY LOCK IN THEORY AND PRACTICE

215. *Remarks Concerning the Safety* Actions.—If we are given the specifications governing the construction of any escapement we can therefrom determine the nature of the safety action.

216. *Specifications Relating to the Safety Lock.*—Our explanations apply to the Elgin type of escapement only. Total lock, 2 degrees. We assume the total lock to be composed of: Drop lock, 1½ degrees; slide lock, ½ degree. Guard-point's freedom from the edge of table roller, with the lever against its bank, 1¼ degrees. Freedom of roller jewel from the slot corner and central part of the horn, 1¼ degrees, with the lever against its bank. Freedom of roller jewel from the end of the horn, with the lever against its bank, 1¾ degrees.

217. *Guard Safety Lock, or Safety Lock Relating to the Guard-Point and Roller.*—The guard-point's freedom from the edge of the roller with the lever against its bank equals 1¼ degrees. The total lock amounts to 2 degrees. If we bring the guard-pin against the edge of the roller we, in consequence, destroy the angle of freedom, viz., 1¼ degrees. Subtracting this from the total lock, we obtain $2° - 1\frac{1}{4}° = \frac{3}{4}°$. The answer means that when we hold the guard-pin in contact with the edge of the roller there still exists a remaining safety lock of ¾ degree. This amount is sufficient to insure the escapement action.

218. *Corner Safety Lock or Safety Lock Relating to the Slot Corner and Roller Jewel.*—The corner of the slot (according to specifications) stands, when the lever rests against its bank, 1¼ degrees from the path of the roller jewel. The total lock is 2 degrees, therefore when the corner of the lever slot is brought into contact with the roller jewel the angle of freedom (1¼ degrees) is destroyed and the lock of the tooth on the pallet is correspondingly lessened, viz., $2° - 1\frac{1}{4}° = \frac{3}{4}°$. Our calculation shows that when the slot corner and roller jewel touch each other the action of the escapement is insured by a safety lock of ¾ degree.

219. *Curve Safety Lock or Safety Lock to the Central Part of the Horn and Roller Jewel.*—In a double-roller escapement the

instant the guard-finger enters the crescent the preservation of the safety action devolves upon the central part of the lever horn and the roller jewel. According to our specifications, when the lever is at rest against its bank the roller jewel will be separated from this part of the horn by a distance equal to 1¼ degrees. The total lock of the tooth on the pallet with the lever against its bank is 2 degrees, therefore when the central part of the horn is brought into contact with the roller jewel the safety lock will equal 2° — 1¼, or ¾°. This safety lock guarantees the action of the escapement.

220. *Separation of the Tip of the Horn from the Roller Jewel.* —The end of the lever horn is, according to specifications, so formed that the path of the roller jewel will pass it at a distance of 1¾ degrees when the lever is at rest against its bank. The freedom of the guard-point from the edge of the roller when the lever rests against its bank equals 1¼ degrees. If we bring the guard-point in *contact* with the roller and guide the roller jewel *opposite* the tip of the horn, the horn and the roller jewel will be separated by a space amounting to ½ degree, viz., 1¾° — 1¼° = ½°.

221. *Detrimental Effect of Erroneously Cutting the Lever's Acting Length.*—We made a plain statement in the earlier part of the preceding lesson—*i. e.*, when angles arise from a common point an alteration of one of the angles is reflected by the remaining angles. This statement we shall now prove. As stated in our specifications, the angle of freedom of the roller jewel with the lever against its bank is 1¼ degrees. The total lock of the tooth on the pallet is 2 degrees. If we cut away the corners of the lever slot so as to provide each corner with 2¼ degrees of freedom from the path of the roller jewel, when the lever rests against its bank, the result would be disastrous to the safety action. This is easily proven from our figures. The lock is 2 degrees, the new freedom of the roller jewel from the slot corner is 2¼ degrees. The freedom *exceeds* the lock. *This is an error*, because if the slot corner is brought into contact with the face of the roller jewel the tooth of the escape wheel would, under test conditions, leave the locking face of the pallet jewel and enter on to the impulse face of the stone, causing a tripping error.

222. *Detrimental Effect of Bending the Guard-Pin.*—If the total lock is 2 degrees and we bend the guard-pin away from the roller so that when the lever rests against its bank the guard-pin is removed 2¼ degrees from the edge of the roller, the effect on the safety action would be ruinous. Subtracting the lock (2 degrees) from the freedom (2¼ degrees) shows that the tooth of the escape wheel would, under test conditions, enter on the impulse face of the pallet jewel; the result would be a tripping error.

LESSON 20

THEORETICAL AND PRACTICAL ANALYSIS OF BANKED TO DROP

223. *Banked to Drop—Its Relation to Drop Lock.*—As previously defined, banked to drop means that the banking pins are turned in to such an extent that slide or second lock is eliminated. Therefore when an escapement is truly banked to drop, we find present only the drop or first lock. The relation of the drop lock to guard and corner freedom we shall now treat on. (Banked to Drop.)

224. *Analysis of the Guard Freedom, Banked to Drop—Elgin Type.*—If we are given the usual figures representing the specifications of an escapement, and desire to make an analysis of that escapement when same is banked to drop, we must deduct the slide from *three* sources—first, from the total lock; second, from the guard freedom; third, from the corner freedom.

Specifications *not* Banked to Drop—Total lock, 2 degrees. Of the total lock, 1½ degrees are drop lock and half a degree of slide. Freedom of guard point from edge of table, lever against bank, 1¼ degrees.

To change the foregoing into banked to drop specifications we deduct the slide from the total lock and also from the guard freedom.

The banked to drop specifications will therefore read:

Drop lock, 1½ degrees.
Guard freedom, ¾ degree.

This means, that when the escapement is banked to drop, and the lever at rest against its bank, that the guard point will be *separated* from edge of table ¾ degree.

It also expresses the fact, that when the guard safety test is used a remaining or safety lock of ¾ degree will be found (1½ — ¾ = ¾).

225. *Analysis of the Corner Freedom Banked to Drop—Elgin Type.*—Specifications *not* banked to drop. Total lock, 2 degrees, composed as follows: Drop lock, 1½ degrees; slide, ½

degree. Freedom of slot corner from path of roller jewel when the lever is at rest against its bank, 1¼ degrees.

When we bank this escapement to drop we deduct the slide from two sources—first, from the freedom of the slot corner with roller jewel (1¼ — ½ = ¾), and also from the total lock (2 — ½ = 1½).

The following now show banked to drop specifications:

Drop lock, 1½ degrees.
Corner freedom, ¾ degrees.

This implies that under banked to drop conditions, with the lever at rest against its bank the slot corner and roller jewel will be ¾ of a degree apart.

When the corner safety test is tried we will find a remaining or safety lock equal to ¾ degree (1½ — ¾ = ¾).

226. *Banked to Drop Summary, Elgin Type*—
Guard freedom present.
Corner freedom present.
Safety lock always less than the drop lock.

227. *Analysis Banked to Drop—South Bend Type.*—Let 2 degrees represent the total lock in an escapement of the South Bend type. Of the total lock 1 degree will represent drop lock and 1 degree slide. The amount of slide always equals the corner and guard freedoms. The extent of drop lock equals the safety lock. When *banked* to drop we will find a drop lock of 1 degree present. As the corner and guard freedoms equal the slide, these freedoms are destroyed by banking to drop.

228. *Banked to Drop Summary—South Bend Type.*—
Guard freedom, none.
Corner freedom, none.
Safety lock *equals* the drop lock.

LESSON 21

THE GUARD TEST IN THEORY AND PRACTICE

229. *Theory of Guard Test.*—The theory underlying the guard test has been partly reviewed in our lesson on the theory of the safety actions. To keep the subject *distinct* we shall again briefly discuss same.

230. *Specifications—Elgin Type.*—Total lock, 2 degrees. The freedom of the guard-point from the roller, 1¼ degrees.

NOTE.—Of the total lock 1½ degrees is the drop lock, the remaining ½ degree being the slide.

231. *Deductions from the Specifications.*—A draft of an escapement made in conformity with the above figures will show a tooth locked on the pallet jewel to the extent of 2 degrees, the lever being at rest against its bank. The freedom of the guard-pin from the edge of the roller will be 1¼ degrees.

232. *Guard Test—Deductions Banked to Drop.*—If we employ the same specifications, but subtract the slide (½ degree), our new specifications will then read: Drop lock, 1½ degrees. The freedom of the guard-point from the roller, ¾ degree.

A drawing made in accordance with these banked-to-drop specifications will show the tooth as locked on the pallet jewel 1½ degrees. The freedom of the guard-point from the edge of the roller will be ¾ degree. The point we desire to impress is that an Elgin type of escapement, when banked to drop, *will always show freedom between the guard-point* and the edge of the roller. This is a fact of great practical importance.

Another fact we wish to be remembered is that an escapement of the South Bend type will *not* show any guard freedom *when banked to drop;* these differences must be kept in mind whenever the guard test is used.

233. *The Guard Test in Practice—Elgin Type.*—Beginners experimenting with this test are advised to bank *every* escapement to drop; accuracy is thereby attained. Assuming we have an escapement before us, the routine of testing the extent of the guard freedom is as follows:

A—The escapement being banked to drop, revolve the balance so as to bring the guard-point opposite the edge of the roller as shown by Fig. 24.

B—Hold the parts in the position indicated by Fig. 24.

C—With a watch oiler reach into the movement and lift the

FIG. 24

lever away from its bank. This brings the guard-pin in contact with the edge of the roller (Fig. 25).

D—The extent the lever can be lifted off its bank represents the freedom of the guard-pin from the edge of the roller.

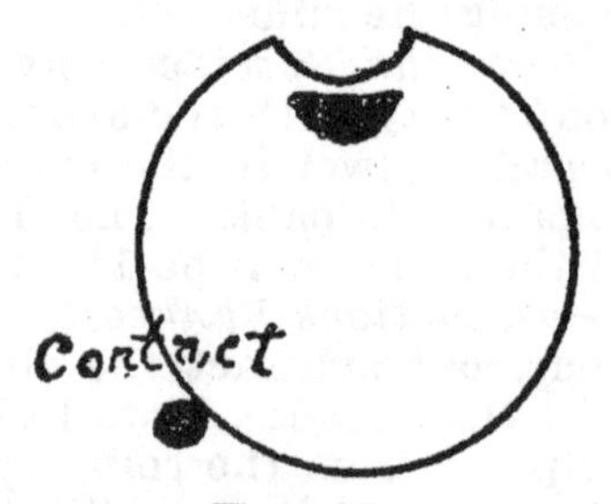

FIG. 25

E—A similar test should be made on the opposite side of the roller. The guard freedoms should be equal.

234. *Some Incorrect Findings Discoverable by Guard Test—Banked to Drop—Elgin Type.*—The subject of incorrect findings discoverable by the guard test is rather too extensive for consideration in this part of the lessons. As some aid we briefly mention the following:

235. *Example A—Elgin Type.*—Should the drop lock in an escapement be correct, namely, neither light nor deep, and the guard test shows *no* freedom, banked to drop, between the guard-point and the roller, the want of guard freedom will indicate that either the guard-point is too far forward or the diameter of the roller is too great.

236. *Example B—Elgin Type.*—If the drop-locks are deep and an excess of freedom is discovered between the guard-point and the roller (banked to drop) we can change the error of excessive guard freedom into a correct guard freedom by lessening the deep lock.

LESSON 22

THE CORNER TEST IN THEORY AND PRACTICE

237. *Theory of the Corner Test.*—Specifications, Elgin type, total lock, 2 degrees. Freedom of the roller jewel from the slot corner, 1¼ degrees.

NOTE.—Of the *total lock*, ½ degree belongs to the slide; the remainder, viz., 1½ degrees, represents the drop-lock.

238. *Deductions in Accordance with Specifications.*—If we make a drawing of an escapement, following the figures in the specifications, the drawing will show the tooth locked on the

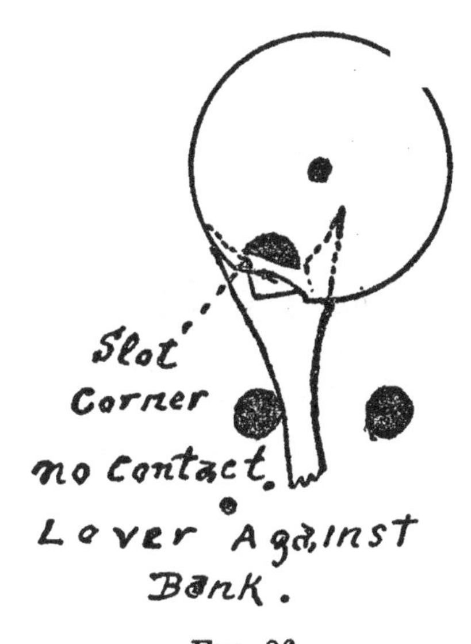

FIG. 26

pallet jewel to the extent of 2 degrees. The lever will be at rest against its bank and the corner of the lever slot will be separated from the path of the roller jewel to the extent of 1¼ degrees.

239. *Banked-to-Drop Specifications—Elgin Type.*—Drop-lock, 1½ degrees. Freedom of the slot corner from the path of the roller jewel, ¾ degree.

NOTE.—The corner freedom *under banked-to-drop conditions* is obtained by subtracting the slide, ½ degree from our first specifications, 1¼ — ½ = ¾.

240. *Corner Test—Deductions When Escapement is Banked to Drop.*—If we follow the specifications, making a drawing therefrom, the lever will be shown as at rest against one banking pin. If the roller jewel is figured in the drawing as opposite the slot corner in the manner shown in Fig. 26 the space separating the slot corner from the roller jewel will equal ¾ degree. The point we wish to emphasize is that an Elgin type of escapement when banked to drop will show, as illustrated in Fig. 26, a little freedom between the slot corner and the roller jewel. If a South Bend type of escapement *is* banked to drop *no* freedom will be discovered between the slot corner and the roller jewel. The difference between the escapement types with regard to their corner and guard freedoms must be remembered and carried into actual practice.

241. *The Corner Test in Practice—Elgin Type.*—To obtain accurate information by the corner test requires that the escapement be banked to drop. The routine of testing the corner freedom is as follows:

A—Bank the escapement to drop and revolve the balance so as to bring the roller jewel opposite the slot corner, as illustrated by Fig. 26.

B—Retain the parts in position as shown in Fig. 26.

C—With a watch oiler or other fine tool lift the lever away

FIG. 27

from its bank. This causes the slot corner to come into contact with the roller jewel, as shown in Fig. 27.

D—The extent we are able to lift the lever away from its bank shows the extent of the corner freedom.

E—Make a test on the opposite slot corner. If conditions are correct the corner freedoms will be equal.

242. *Example A—Elgin Type.*—When the drop-locks are *correct* and the corner test shows *no* freedom between the slot corners and roller jewel, as shown in Fig. 28, we realize that

FIG. 28

either the lever's acting length is long or the roller jewel's position is too far advanced.

243. *Example B—Elgin Type.*—Should the drop-locks be deep and the corner test show an excess of corner freedom the error of excessive corner freedom can be overcome and corrected by simply decreasing the drop-locks.

LESSON 23

SLIDE—ITS RELATION TO CORNER AND GUARD FREEDOM—ELGIN AND SOUTH BEND TYPES

Remarks.—Although "slide" and "freedom" have been discussed in Lesson 20 and elsewhere, we have, because of their importance, given them further and separate consideration.

244. *Slide, and the Provision for Corner Freedom—Elgin Type.*—If we accept 2 degrees as the *total lock* belonging to an Elgin type of escapement, and of this total allow 1½ degrees for the drop-lock, the remaining ½ degree will be slide. In paragraph No. 207 the freedom of the roller jewel from the slot corner, with the lever against its bank, escapement *not* banked to drop, is given as 1¼ degrees. These figures therefore represent the amount of corner freedom when slide is present.

To estimate the corner freedom when this type of escapement *is banked to drop* we deduct the slide ½ degree from the above corner freedom of 1¼ degrees. Therefore when banked to drop the corner freedom equals ¾ degree (1¼° — ½° = ¾°).

245. *Slide, and the Provision for Guard Freedom—Elgin Type.*—Specifications—Total lock, 2 degrees, composed as follows: Drop lock, 1½ degrees; slide, ½ degree. Freedom of guard point from edge of table, 1¼ degree, when lever is at rest against its bank and escapement is *not* banked to drop.

To determine the guard freedom when escapement *is banked to drop* subtract the slide ½ degree from the original guard freerom, 1¼ degrees; this leaves ¾ degree as amount of guard freedom when banked to drop.

246. *Summary of Corner and Guard Freedoms—Elgin Type.*—

Slide Present—*Not* banked to drop corner freedom, 1¼ degrees; *not* banked to drop guard freedom, 1¼ degrees.

Slide Absent—*When* banked to drop corner freedom, ¾ degrees; *when* banked to drop guard freedom, ¾ degrees.

The lesson to be learned from above is, that an Elgin type of escapement *when banked to drop* shows *both* guard and corner freedoms.

When slide is present the freedoms are increased by exactly the amount of slide. Compare with No. 249.

To protect the safety locks, the drop lock in an Elgin type of escapement always exceeds the corner and guard freedoms.

247. *Slide and Corner Freedom—South Bend Type.*—Specifications—Total lock, 2 degrees. The total lock is made up of: Drop lock, 1 degree, and slide, 1 degree. Freedom of slot corner from roller jewel, lever against its bank, 1 degree.

According to the foregoing figures, a South Bend type of escapement possesses a corner freedom of the same amount as the slide. Therefore, when *not* banked to drop, we find *the same* amount of slide and corner freedom.

When this escapement *is banked to drop* we of course thereby destroy the slide. As the amount of slide equals the amount of corner freedom, *no* corner freedom will be found when banked to drop.

248. *Slide and Guard Freedom—South Bend Type.*—Specifications—Total lock, 2 degrees; of this amount, 1 degree will represent drop lock and 1 degree slide. The freedom of the guard point from edge of table, lever against bank is to be 1 degree.

The specifications given show, that a South Bend type of escapement possesses slide equal in amount to guard freedom. By banking this escapement to drop we remove both slide and guard freedom.

249. *Summary Corner and Guard Freedoms—South Bend Types.*—

Slide Present—*Not* banked to drop corner freedom, 1 degree; *not* banked to drop guard freedom, 1 degree.

Slide Absent—*When* banked to drop corner freedom, none; *when* banked to drop guard freedom, none.

As shown by the summary when slide is present, both guard and corner freedom are present. When slide is absent in a South Bend type of escapement the freedoms are likewise absent. See No. 246 for comparison.

LESSON 24

THEORY AND EXPLANATION OF THE ANGU LAR TEST—ELGIN TYPE

250. *Uses of the Angular Test.*—The angular test is of all tests the most accurate for determining if an escapement is either *in* or *out* of angle. It is the dominant test for yielding information relative to the action of the roller jewel as it relates to the fork. It expresses as no other test can or does the close association which exists between the *lock* (drop lock) and the *length* of the lever. We shall state the principles governing the test, making use of specifications for this purpose.

251. *Specifications.*—Escapement banked to drop, Elgin type: Drop-lock, 1½ degrees; lift on tooth, 3 degrees; lift on pallet, 5½ degrees; freedom of the roller jewel from the slot corner, 1 degree (banked to drop).

By adding the drop-lock and both of the lifts together we obtain the lever's angular motion from bank to bank as 10 degrees. If we have an escapement built in conformity yith the above specifications, the same being banked to drop, we would on using the angular test learn what is meant by the correct relationship of the roller jewel-fork action to the tooth and pallet action. In other words, the angular test will decisively inform us if the parts are well matched. We can also with exceptional accuracy determine if an escapement is in or out of angle.

252. *Preliminary Explanation of the Angular Test—Elgin Type.*—In accordance with instructions supplied in a later lesson on the angular test we assume that the motion of the lever is blocked, viz., wedged. Then, by placing a finger on the balance rim, cause the balance to rotate, thereby bringing the roller jewel into the slot. This rotation is continued until the roller jewel pushes its way past the slot corner. If we then make an examination of the escapement parts two facts will be noticed: First, the lever fails to reach its bank, although acting under banked-to-drop conditions. Secondly, the tooth of the escape wheel *remains in contact* with the pallet jewel after the manner shown in Fig. 29.

253. *Theoretical Explanation of the Angular Test—Elgin Type.*—Why the lever *failed* to reach its bank and why the tooth *remained in contact* with the pallet jewel is best explained by means of the specifications. The specifications (banked to drop) call for 1 degree of freedom between the roller jewel and the corner of the slot when the lever is at rest against its bank. Under test conditions (lever wedged) the space we see separating the side of the lever from the banking pin is the equivalent of the stated freedom of the roller jewel, from the slot corner as given in the specifications. This amounts to 1 degree. The lever therefore fails to reach the opposite bank by 1 degree of angular motion.

We figured in the specifications that the lever's angular motion from bank to bank equals 10 degrees. The contact of the roller jewel with the fork slot amounts to 9 degrees (10° — 1° = 9°). Therefore 9 degrees represents the actual amount traveled by the lever. As it takes 10 degrees of angular motion to release a tooth from the pallet, it can now be plainly understood why the tooth (lever wedged) must remain in contact with

FIG. 29

the pallet jewel, as illustrated in Fig. 29. The facts above stated apply to all escapements of the Elgin type, irrespective of their specifications.

254. *Summary of Angular Test Findings—Elgin Type.*—These facts we shall now briefly repeat. An escapement of the Elgin type when the parts involved are correctly matched will show by the angular test:

A. That the lever is *unable* to reach its opposite bank.

B. That the teeth of the escape wheel will *remain in contact* with the pallet jewel's, as shown in Fig. 29.

LESSON 25

THEORY AND EXPLANATION OF THE ANGULAR TEST—SOUTH BEND TYPE

255. *Specifications—Angular Test—South Bend Types.*—Drop lock, 1 degree; lift on pallet, 4½ degrees; lift on tooth, 3½ degrees.

The lever's angular motion is the sum of the above, viz., 9 degrees.

The above specifications indicate, because of the absence of slide, that the escapement *is* banked to drop.

256. *Theoretical Explanation of the Angular Test—South Bend Type.*—The lever's angular motion is composed of the drop-lock, the lift on the tooth and the lift on the pallet; their total equals 9 degrees. As there is *no* provision in the specifications for the freedom of the roller jewel from the slot corners (corner freedom) *under banked-to-drop conditions*, we must therefore expect to find *contact* of the roller jewel with the fork slot during the lever's motion from bank to bank *when subjected to the angular test.* To avoid confusing beginners we wish to point out that when the bankings are opened for slide the necessary corner freedom is thereby provided.

257. *Summary of Angular Test Findings—Banked to Drop—South Bend Type.*—Given an escapement constructed according to the specifications before stated, it will be found that when the angular test is used and the escapement is *banked to drop* the following takes place:

A—The roller jewel will *touch* the slot corners either passing in or out of the slot.

B—The lever will move from one bank to a position of *contact* with the opposite bank.

C—The escape-wheel teeth will be *discharged* from each pallet jewel.

These findings are contrary to the proof findings of an Elgin type of escapement. The difference between the types must be remembered.

LESSON 26

THE ANGULAR TEST IN PRACTICE—VARIATIONS, AND OUT OF ANGLE

258. *How to Apply the Angular Test.*—Commence by banking the escapement to drop. The escapement being banked to drop, place a wedge under the lever. The material used for a wedge varies with the accessibility of the lever. For use in full plate watches the wedge can be made from a piece of very weak and narrow mainspring. The piece selected should be about one inch in length. For convenience and practical reasons form the wedge into the shape of a bow. This bow-shaped wedge is especially useful in full plate watches. When inserted under the lever it acts as a spring cushion in retarding the lever's motion. The suggested steel wedge cannot be used in all watches owing to recesses in plate, in which event cork or pith can be employed. Having the lever wedged and the balance in position, place a finger on the balance rim and start revolving same, thereby bringing the roller jewel into the slot and out the opposite side. As a matter of precaution it is wise to *cease turning* the balance, when in passing out, the center of the roller jewel comes opposite the slot corner. This is advised to avoid any chance of the guard-point engaging the edge of the roller and thereby falsely increasing the lever's motion. The roller jewel being moved into the desired position, remove the finger from the balance, then with an eye-glass observe the relation of the tooth and pallet jewel. If the lock is correct and the lever's acting length likewise correct the tooth will be found barely in contact with the releasing corner of the pallet jewel, as shown in Fig. 29. A test on the opposite side should reveal like contact of tooth and pallet. When the contact of each tooth with its pallet jewel resembles Fig. 29 we know that the parts involved are well matched and the escapement correct from the standpoint of the angular test's proof-findings. Proof-findings when discovered teach us that the roller jewel's action with the fork is exactly adapted to the drop lock. The angular test will at times show tooth and pallet conditions not in conformity with the proof-findings. When irregular conditions are found, examine the

nature and extent of drop-lock and make use of the corner test to assist in uncovering the cause of error. The proof-findings above mentioned refer only to an Elgin type of escapement.

259. *The Guard-Point and the Angular Test.*—When using the angular test it is advisable to remove the guard-point from proximity to the roller. Should the guard-point be too close to the roller it may, when the lever is wedged, touch the roller's edge, thereby causing an increased motion of the lever. This motion would alter the position of the tooth with relation to its pallet. This would lead to an erroneous decision.

260. *Blocking the Lever—Angular Test.*—Blocking the lever's motion is a quick way of applying the angular test. Its use is not advisable for beginners, because some previous experience by the slower method of wedging is necessary to prevent erroneous deductions. Blocking the lever possesses this advantage: An escapement can be examined in its original condition, namely, with the guard-pin straight and slide present. Should errors be discovered, then, to make certain the guard-point must be removed from the edge of the roller and a retest made. The following is the routine:

First—With the lever at rest against its bank.

Second—Take a watch oiler and press it against the side of the lever, apparently with the intention of retaining the lever against its bank.

Third—Place a finger on the balance and guide the roller jewel into the slot.

Fourth—Hold the tool against the side of the lever and cause the roller jewel *to push* the lever toward the opposite bank.

Fifth—Keep turning the balance until the roller jewel is "felt" *to just emerge* from the slot.

Sixth—Carefully *retain* the tool against the side of the lever in the *exact* position it occupied when the roller jewel—see No. 5 —started to scrape its way past the slot corner and examine the relation of tooth and pallet.

261. *Variations from Proof Findings of the Angular Test.*—In actual bench practice many variations from the standard proof findings will be encountered. For instance a tooth on one pallet may show more contact than another tooth on the opposite pallet; or one tooth may remain in contact, the other tooth being discharged, etc. Conditions departing from normal proof findings express the fact that some escapement error is present. Beginners in escapement testing should learn to locate the error by means of banking to drop the guard, corner and safety tests. Knowing these, no trouble will be experienced in reading departures from the proof findings. We know from actual experience with young watchmakers that, given a knowledge of the nature of lock, banking to drop, corner, guard and safety tests, they can

successfully attempt escapement alterations. Especially so when alterations are checked by the proof findings of the angular test.

262. *Out of Angle, as Shown by the Angular Test—Elgin Type.*—As previously mentioned, *the proof findings* of an Elgin type of escapement will show similar points of contact of each tooth with its pallet jewel. Out of angle conditions are expressed as follows:

A. Any departure *from corresponding positions of contact* declares the escapement as out of angle.

B. If the amount of contact of one tooth with its pallet jewel *exceeds* the amount of contact shown by the tooth on the opposite pallet the escapement is out of angle.

C. When one tooth *remains* in contact and the other tooth is *discharged* from its pallet jewel the escapement is out of angle.

263. *Causes Producing Out of Angle.*—

A. The most frequent cause of an escapement being out of angle is due to irregularities in the drop locks; that is, the drop lock on one pallet differs in extent from the drop lock on the opposite pallet. Of course equalizing the drop locks corrects "out of angle."

B. Out of angle is sometimes attributable to the lever being bent.

C. A bent lever and inequalities of the drop locks combined may also cause an escapement to be out of angle. Whatever the source of this error may be, it generally requires correction.

264. *Out of Angle—South Bend Escapement Types and the Angular Test.*—When testing to learn if an escapement of the South Bend type is out of angle, *do not* bank the escapement to drop. Theoretical explanations already given, together with practical experiments will establish the reason for this statement. Prior to investigating the subject of out of angle in escapements of the South Bend type, students should become familiar with "out of angle" as found in escapements of the Elgin class. No trouble will then be experienced in detecting "out of angle" in South Bend escapements.

LESSON 27

DROP LOCK—ITS VARIATION

265. *Drop Lock—As the Watchmaker Finds It.*—That the expression "*correct drop lock*" means drop lock of varying quantity is a fact apparent to every practical watchmaker. A drop lock that is suited to a watch of high grade would be unsuitable for a low-grade watch. This statement is not made from a theoretical viewpoint, but from the standpoint of the man at the bench. Writers have a habit of ignoring all conditions of lock save one—that is, drop lock in its theoretical form. Nevertheless the repairer is constantly encountering forms of drop lock differing widely from the theoretical variety. The watchmaker in his daily work meets with escapements as they are—practical—and rarely as they ought to be—theoretical. This intermixture of escapement construction is puzzling until read in the light of the angular test.

266. *Three Types of Escapements.*—The man at the bench encounters three types of escapement perfection. Each in its way gives satisfaction, because the associated parts are well matched and therefore suited to each other. First, we have the perfect escapement, which is rather rare; second, the correct escapement, frequently found in high-grade watches; third, the more common and plentiful type—namely, the commercially correct escapement. The governing feature of each escapement type is that the parts are well matched. The parts of the *perfect escapement* are matched theoretically and practically, consequently an exact harmony of action prevails.

In the *correct escapement* the parts are also well matched, but at the expense of an increased frictional resistance.

The parts of the *commercially correct escapement* are likewise matched, a fact capable of proof by the angular and other tests.

Students must learn to recognize the distinctions named and to get away from the misleading idea that many escapements, because the parts are matched, are theoretically perfect. The investigating student will soon learn that the majority of escapements are *commercially matched*, the minority being matched theoretically.

267. *Escapement Matching.*—By correct or commercial matching of an escapement we refer particularly to a drop lock *exactly suited* to a given length of lever and roller jewel radius. The fact is that in the sense mentioned escapement tests may agree in declaring the parts matched—that is, the drop lock is adapted to the action of the roller jewel with the fork. This, however, does not imply escapement perfection, as our future test lessons and experiments will prove.

268. *Drop Lock in a Perfect Escapement.*—If an escapement of the Elgin type possesses a theoretically correct drop lock, and associated with this is a lever whose acting length is theoretically correct, the parts are assuredly well matched. The angular test will express this fact by showing the correct amount of contact of the tooth with the pallet, as illustrated in Fig. 29. This drawing represents the proof-findings.

The corner test will also demonstrate that the correct amount of corner freedom is present.

269. *Drop Lock in a Correct Escapement.*—An escapement of this class—Elgin type—possesses a drop lock somewhat greater than that found in a theoretical or perfect escapement. When a "correct escapement" possesses a lever whose acting length and roller-jewel radius are *adapted* to the amount of drop lock present the angular test will show contact of tooth and pallet as represented in Fig. 29.

The corner test will also show a corner freedom of suitable amount.

In a "correct escapement" we also find there exists a smooth, concerted action of the escapement parts as revealed by the tests. In other words, the parts are well matched.

270. *Drop Lock in a Commercialy Correct Escapement.*—The majority of escapements belong to the commercialy matched class. Such escapements have a drop lock *deeper* than that found in the "correct escapement" mentioned previously. Associated with the *greater lock* will be a lever or roller-jewel radius of *increased length*, but *exactly suited* to the increased drop lock. When escapements of the "commercially correct" variety are provided with either a lever whose acting length matches the drop lock or a roller jewel whose radius corresponds with this lock the angular test will show tooth and pallet contact as illustrated in Fig. 29.

The corner test will show the escapement as provided with a corner freedom exactly suited to its condition, but exceeding in amount the corner freedoms of the two preceding types.

That the parts are matched is as much in evidence in the "commercial" as in the higher escapement types.

It should be the aim of every watchmaker to at least change the more commercial products into a more perfect type of escapement. Usually the change can be made with but little trouble once the primary principles of the locks and tests have been mastered.

At first it is advisable for students to confine their studies and experiments to the Elgin type. Dueber and South Bend escapements will afterward present no difficulties.

LESSON 28

BANKING TO DROP IN PRACTICE

271. *Findings—Banked to drop.*—The student ambitious to advance *must* experiment with escapements *banked to drop.* The findings stated below will be met with in all watches *when the parts are well matched.* These test findings represent the proof or correct findings—banked to drop—for escapements of the Elgin and South Bend types. A few practical experiments will show that many departures from our test standards prevail, the irregularities being due to escapement faults. Full instruction regarding the detection and correction of errors will be found in the pages covered by the "Test Lessons."

A study of these lessons will lead the student out of various escapement difficulties.

272. *Banked to Drop Findings (Elgin Type — Parts Matched).*—

Guard Test—*A slight freedom* between the guard point and roller—guard freedom. (See drawing No. 24.)

Corner Test—*A slight freedom* between the slot corners and roller jewel—corner freedom. (See Fig. 26.)

Angular Test—*Each* tooth *shows contact* with its pallet jewel as illustrated in Fig. 29.

Contrast the above Elgin findings with the following, representing findings iu the South Bend type. Then by *practical experiments* impress on your memory the ever-useful and important truths stated in this and the following paragraph.

273. *Banked to Drop Findings (South Bend Type—Parts Matched).*—

Guard Test—*No* freedom between the guard point and roller table.

Corner Test—*No* freedom between the slot corners and the roller jewel.

Angular Test—*No* contact of tooth and pallet, as illustrated in Fig. 30.

LESSON 29

TYPES OF AMERICAN ESCAPEMENTS

274. *American Escapement Types.*—Two different types of escapements are found in American-made watches. For convenience we have designated them Elgin and South Bend. The term "Elgin type" applies to *all* makes, *excepting* South Bend and Dueber, the latter being associate escapement types. The difference in type is extreme as viewed from the standpoint of our tests, a fact readily discovered by comparing the banked-to-drop test findings of each type as already set forth in Lesson 28.

275. *Lock Division of the Elgin Type.*—Of the total lock in an escapement of the Elgin type—in round numbers—practically two-thirds represent the drop lock, the remaining one-third being slide.

276. *Lock Division of the South Bend Type.*—The total lock in an escapement of the South Bend type is equally divided—viz., one-half represents the drop lock, the other half is slide.

Note on Foreign Escapements—As a general rule, examine a foreign-built escapement as you would an Elgin. While this may not be an invariable rule, yet, so far as the writer knows, the Elgin is the most applicable type for this purpose.

LESSON 30

ESCAPEMENT EXAMINATION

277. *Preliminary Advice.*—The tests stated in the lessons are accurate and reliable, because they are founded upon both a theoretical and practical basis. No matter how accurate tests may be, mistaken deductions will be made unless the below instructions are closely followed out.

278. *General Preparatory Instructions.*—

(a) See that all hole jewels are *tight* in their settings, and that jewel settings are tight in their seats.

(b) See that all pivots correctly fit their respective holes. Should defects in fitting be discovered, make the necessary corrections.

(c) Attend to the end shakes of all parts.

(d) The hair spring requires to be true, level, and correctly centered. Free from the balance arm and bridge, and tightly pinned at collet and stud.

(e) The banking pins must be tight and upright.

(f) The guard point must be tight and correct in shape.

(g) See that the lever and its attached parts are secure.

(h) The roller table should be true in flat and round. Tight on staff and with smooth edges.

(i) The condition of "draw" must be investigated.

(j) The extent of each drop lock requires to be known. An estimate of the degrees of each lock should be made.

(k) The number of degrees of slide should likewise be known.

(l) The inside and outside drop should be examined, and, if necessary, the extent of the drop in degrees determined.

(m) Shake, "inside and out," calls for careful examination.

(n) Examine the lifts of tooth and pallet. *Especially observe their manner of engaging and disengaging.*

(o) Determine amounts of "guard freedom," banked and *not* banked to drop.

(p) Investigate the condition of the "corner freedoms," banked and *not* banked to drop.

(q) Employ the safety tests to try out all the safety locks.

(r) Make use of the angular test and learn if the parts are matched or otherwise.

279. *Routine Escapement Examination.*—The following is a short outline of an escapement examination:

First—Bank the escapement to drop.

Second—Inspect the drop lock.

Third—Inspect the inside and outside drops.

Fourth—Inspect the shakes inside and out.

Fifth—Inspect the draw on both pallets.

Sixth—As a precautionary measure remove the guard point from the edge of the roller—that is, increase the distance which normally separates these parts.

Seventh—Try the corner test.

Eighth—Try the corner safety test.

Ninth—Make use of the angular test.

(Here we pause to make the necessary alterations in accordance with directions given in the "Test Lessons.")

Tenth—Readjust the guard point to the roller.

Eleventh—Try the guard test.

Twelfth—Try the guard safety test.

As the student becomes familiar with escapement work and the tests the routine of examination can be much shortened, as shown by our "Bench Problems," examples being given in later lessons.

LESSON 31

SUMMARY OF THE TESTS

280. *Divisions of Tests.*—The division and subdivision of the tests as gathered together in this lesson are given in a form convenient for reference and bench usage. The tests for freedoms in the Elgin type *banked to drop are*, respectively, the guard test, the corner test and the curve test.

The subdivisions of these tests are the guard safety test, the corner safety test, and the curve safety test. Their distinction, difference and application must be grasped by every student. The best way to learn the tests is the practical way—namely, with a watch in hand. The angular test is not included in this summary, full details having been given in a preceding lesson.

281. *Testing the Draw.*—

First—Remove the balance.

Second—Lift the lever off its bank with a piece of pegwood, but not sufficient to cause unlocking.

Third—Remove pegwood.

Fourth—If the draw is sound the lever immediately returns to its bank. If the escapement is freshly oiled and clean and the draw on either pallet proves defective a correction is necessary. (See Test Lessons on Draw.)

282. *Testing the Lock.*—The locks should be tested both by observation and the angular test. Should either the drop lock or the slide lock be found defective a correction is necessary. When in doubt about the extent of drop lock bank the escapement to drop, then retest.

An estimate of the amount of drop lock, in degrees, can be made by aid of the tables given in paragraph No. 180.

Note.—When the locks are tested, the student is advised to *carefully observe the "lifts,"* as directed in Lesson 11.

283. *Testing the Inside Drop.*—

First—To make this test students are advised to bank the escapement to drop. When experience is gained this is not necessary.

Second—Cause a tooth to be discharged from the letting-off corner of the *entering* pallet and note the amount of its drop. (See Lesson 13, paragraph No. 184.)

284. *Testing the Outside Drop.*—

First—Bank the escapement to drop.

Second—Cause the discharge of a tooth from the releasing corner of the exit pallet and observe the extent of drop. (Consult Lesson 13, Table of Drops, paragraph No. 184.)

Inside and outside drop should be equal. The degrees of drop can be estimated as explained in Lesson 13.

285. *Testing the Inside Shake.*—

First—Under the lever bar place a wedge of tissue paper.

Second—Allow a tooth to drop on the locking face of the *exit* pallet jewel.

Third—Bring the tooth at rest on the *exit* pallet's locking face down to the lowest locking corner of this pallet.

Fourth—Note the space separating the *back of the receiving* pallet from the heel of the tooth just behind it. The space observed between the back of the pallet and heel of the tooth is the inside shake.

286. *Testing the Outside Shake.*—

First—Place a wedge under the lever bar.

Second—Allow a tooth to drop on the locking face of the *receiving* pallet stone.

Third—Move the lever so as to bring the tooth at rest on the locking face of the *receiving* pallet down to the pallet's lowest locking corner.

Fourth—Then observe the space separating the *back* of the exit pallet from the heel of the tooth just behind it. This space is the outside shake, and is the position of least freedom of the parts concerned. The inside shake and outside shake should be equal.

287. *Testing the Freedom of the Roller Jewel in the Slot.*—

First—Place a wedge under the balance rim.

Second—Rotate the balance so that the roller jewel stands centrally in the slot; the lever will then stand midway between the bankings.

Third—With the aid of a tool find out, by shaking the lever, how much side play the roller jewel has when within the slot.

288. *Guard Test Findings—Single and Double Roller.*—The purpose of the guard test is to learn the relation of the guard point to the roller. When an escapement of the Elgin type *is banked to drop* a slight space or guard freedom should be found. When a Dueber or South Bend escapement *is banked to drop* the correct finding by the guard test is that of *contact* of the guard point with the edge of the roller.

289. *Guard Safety Test Findings, Single and Double Roller—Tripping Test.*—The intention of the guard safety test is to determine the condition of the safety lock. If the safety or remaining lock is absent a tripping error is present. This or any of the safety tests can be made with or without banking the escapement to drop.

METHOD OF MAKING THE GUARD TEST—SINGLE AND DOUBLE ROLLER

290. *Guard Test—Elgin Type—*

First—Bank the escapement to drop.

Second—Lift the lever off its bank.

Third—The effect of our second operation is to bring the guard point in contact with the edge of the roller.

Fourth—The extent we are able to lift the lever away from its bank represents the guard freedom.

291. *Guard Test—South Bend Type.—*

First—Bank the escapement to drop.

Second—The lever *cannot* be lifted off its bank because the parts under consideration will *touch.* Consequently the guard freedom is and should be absent

METHOD OF MAKING THE GUARD SAFETY TEST—SINGLE AND DOUBLE ROLLER

292. *Guard Safety Test—Tripping Test.*

First—Bring the guard point in contact with the edge of the roller.

Second—Hold the guard point in contact with the edge of the roller.

Third—While the parts are held in touch with each other, with an eyeglass observe the extent of the remaining or safety lock of the tooth on the pallet jewel's locking face. If there is no safety lock the tooth will enter on to the pallet jewel's impulse face. This means a tripping error is present. Its cause and correction will be found in that part of the book relating to this subject. (See Test Lesson on the Guard Safety Test.)

293. *Corner Test Findings—Single and Double Roller.*—By means of the corner test we learn the relation of the slot corners to the roller jewel. This should be done with the escapement banked to drop. An Elgin type of escapement, when banked to drop, will show by the corner test *a little* freedom—corner freedom—between the slot corner and the roller jewel. A Dueber or South Bend escapement when banked to drop shows *no* freedom between the slot corners and the roller jewel.

294. *Corner Safety Test Findings, Single and Double Roller—Tripping.*—The purpose of the corner safety test is to investigate

the safety lock. A safety lock is a necessity and, if lacking, must be provided as directed in the test lesson.

METHOD OF MAKING THE CORNER TEST—SINGLE AND DOUBLE ROLLER.

295. *Corner Test—Elgin Type.—*

First—*Bank the escapement to drop.*

Second—Rotate the balance so as to bring the roller jewel opposite the corner of the lever-slot.

Third—By means of some fine tool lift the lever off its bank, thereby causing the slot corner to touch the roller jewel.

Fourth—The extent we are able to lift the lever from its bank represents the amount of corner freedom.

296. *Corner Test—South Bend Type—*

First—*Bank to drop.*

Second—The roller jewel when passing in or out of the slot will *touch* the corners. This means corner freedom is *not* present, which finding is correct.

METHOD OF MAKING THE CORNER SAFETY TEST—SINGLE AND DOUBLE ROLLER

297. *Corner Safety Test—Tripping Test.—*

First—Rotate the balance so as to bring the face of the roller jewel opposite the corner of the lever slot.

Second—Hold the roller jewel in the position mentioned above.

Third—With a fine tool lift the lever away from its bank, thereby causing the corner of the slot to come in contact with the roller jewel.

Fourth—Hold the parts in contact and with an eyeglass; inspect amount of the remaining or safety lock. An absence of safety lock indicates a tripping error, which must be corrected.

METHOD OF MAKING THE CURVE TEST—SINGLE AND DOUBLE ROLLER

298. *Curve Test—Single Roller—*

First—With a watch oiler or other fine tool *hold* the guard pin against the edge of the roller.

Second—Rotate the balance, thereby bringing the roller jewel past the horn and into the slot.

Third—While roller jewel is passing the horn, contact being maintained, no contact of the upper part of the horn with the roller jewel should be detected.

Fourth—Immediately the guard pin *enters* the crescent the roller jewel and horn will come in contact. A slight friction of the roller jewel with a small part of the horn and with the slot corner will be detected, but nothing resembling a catch is allowable. The extent of the horn with which the roller jewel under test conditions can come into contact depends upon the width of the crescent.

299. *Curve Safety Test, Single Roller—Tripping Test.*—As mentioned in the curve test, once the guard pin enters the crescent a small part of the horn can be brought into contact with the roller jewel. Immediately this happens *retain the parts in contact* and examine the condition of the safety lock. As a matter of fact, the curve safety test is of little importance in single-roller escapements. In double-roller escapements it is a very important test.

300. *Curve Test—Double Roller.*—

First—Lift the lever away from its bank, thereby we bring the guard finger and roller jewel in contact.

Second—Maintain the parts in contact and revolve the balance, so as to bring the roller jewel past the end of the lever horn and in the direction of the slot.

Third—Continue to slowly revolve the balance, at the same time keeping the guard finger pressed against the table's edge. The instant the guard finger enters the crescent, the roller jewel and curve of horn come into contact. This contact continues until the roller jewel enters the slot. The roller jewel should slide over the face of the horn and into the slot without developing any undue friction.

301. *Curve Safety Test, Double Roller—Tripping Test.*—The fourth section of the foregoing test says, "Immediately the guard pin enters the crescent the roller jewel and horn will come in contact." When the parts stand in the position quoted, then to examine the safety lock hold the horn and roller jewel in touch with each other while the extent of the remaining or safety lock is investigated. A want of safety lock means a tripping error. This fault must be corrected and alterations made in accordance with the directions given in that part of the book treating on Bench Problems and Test Lessons. Students who do not quite understand any particular point or subject will find the series of "Questions" a most useful feature.

BUTTING ERROR

301 A. *Test for Butting Error.*—The student is referred to Lesson 58.

LESSON 32

CLASSIFYING DROP LOCK—CORNER FREE DOM AND GUARD FREEDOM

302. *Drop Lock Classification.*—In the test lessons drop lock is classified into its three main forms, namely (see 265), correct, light and deep.

To classify the amount of drop lock as correct, light or deep, for any given escapement, combinedly requires the employment of the angular test, and an estimate of the extent of lock as directed in paragraph No. 183. Lesson 27 should also be consulted.

303. *Corner Freedom Classification.*—Corner freedom, when the escapement is of the Elgin type and banked to drop, is divided into correct freedom, excessive and contact or no freedom.

The "Test Lessons" specify the corner freedom for an Elgin type of escapement when banked to drop, as correct, excessive, or wanting. Of the three, the most important to learn to recognize is that which we have designated "correct freedom." This, like its intimate associate, "correct lock," is a varying quantity. The correct amount of corner freedom as found in high-grade watches differs (from the standpoint of the man at the bench) from the correct amount of corner freedom associated with low-grade watches; yet each in its place is correct for that escapement of which it is a part. An observation and estimation of the extent of drop lock, and a few experiments with the corner and angular tests will place the student in a position to classify the freedoms.

304. *Guard Freedom Classification.*—In an Elgin type of escapement, when banked to drop, the guard freedom will be found as follows: Correct, excessive, or wanting. One of the three distinctions will describe conditions present:

Correct guard freedom, like its predecessors, correct lock and correct corner freedom, varies with the grade of escapement. Correct guard freedom, as discovered in a low-grade watch, would be unsuited to a high-grade escapement.

The term correct guard freedom is and will be found elastic. In quantity it should about equal the corner freedom. Its extent should not endanger the safety lock nor cause a butting error.

305. *Classification as Used in the Test Lessons.*—Regarding the "Test Lessons," in the following pages, the observed routine is that of treating and discussing *one* error, and teaching the relationship of this error to *three* different kinds of drop lock, viz.: Correct, light, and deep.

As an example of method pursued, Test Lesson 3 A treats on the error of *excessive* corner freedom when associated with *correct drop lock.*

Test Lesson 3 B likewise treats on the error of *excessive* corner freedom, including the added error of *light drop lock.*

Test Lesson 3 C again discusses the error of an *excess* of corner freedom combined with the error of *deep lock.*

In this manner the test lessons are linked together and prominent escapement truths are plainly set before the student.

LESSON 33

RULES GOVERNING ALTERATIONS

306. *Advice and Remarks.*—The student must acquire a thorough knowledge of the general effects caused by any contemplated escapement alteration. For instance, making the drop locks deeper will produce the "effects" mentioned in Rule 1. In a practical way study and work out each rule. By doing so you will find it an easy matter to forecast the result of an alteration. Don't neglect the Questions on Rules and Alterations. They will be found helpful in impressing the practical facts contained in the rules.

Students should at first confine their practical experiments to escapements of the Elgin type, *always keeping same banked to drop.*

Once you become familiar with the principles governing alterations in this type, no trouble will be experienced in adapting the information to escapements of the South Bend type.

Young watchmakers, when undirected or but poorly instructed, find the escapement a hard road to travel. Such will find the Tests, Rules and Test Lessons lamps on the road to practical efficiency. Its "up to you," watchmaker or student, by reading and practical experiment, to get out of the rut.

Experienced workmen desiring only a knowledge of the tests, etc., are referred to the following paragraphs: 265 to 301; 379 to 384; 209 to 264; 306 to 323.

Index to Rules

307. *Index to Rules.*—

General Rules

RULE 1.

308. *Increasing Drop Lock—*

ALTERATION—*Making the drop locks deeper causes the following:*

Effect A—The bankings are spread further apart. (Banked to drop.)

Effect B—Increases the guard freedoms.

Effect C—Increases the corner freedoms.

Effect D—Increases safety locks.

Effect E—Alters both drop and shake. (See below.)

NOTES ON RULE 1

Receiving Pallet Jewel

ALTERATION—*Drawing out only the receiving stone.*

First Effect—Increases drop lock on receiving stone.

Second Effect—Increases drop lock on discharging stone.

Third Effect—Increases inside drop.

Fourth Effect—Increases inside shake.

Dicharging Pallet Jewel

ALTERATION—*Drawing out only the discharging stone.*

First Effect—Increases drop lock on discharging stone.

Second Effect—Increases drop lock on receiving stone.

Third Effect—Increases outside drop.

Fourth Effect—Increases outside shake.

RULE 2

309. *Decreasing Drop Lock.—*

ALTERATION—*Making the drop locks lighter produces the following effects:*

Effect A—The bankings are brought closer together. (Banked to drop.)
Effect B—Decreases the guard freedoms.
Effect C—Decreases the corner freedoms.
Effect D—Decreases safety lock.
Effect E—Alters both drop and shake. (See below.)

NOTES ON RULE 2

Receiving Pallet Jewel

ALTERATION—*Pushing back receiving stone only.*
First Effect—Decreases drop lock on receiving stone.
Second Effect—Decreases drop lock on *discharging* stone.
Third Effect—Decreases inside drop.
Fourth Effect—Decreases inside shake.

Discharging Pallet Jewel

ALTERATION—*Pushing back discharging stone only.*
First Effect—Decreases drop lock on discharging stone.
Second Effect—Decreases drop lock on receiving stone.
Third Effect—Decreases outside drop.
Fourth Effect—Decreases outside shake.

RULE 3

310. *Increasing Corner Freedom.*—
ALTERATION No. 1—*Decreasing the lever's acting length, or roller jewel radius.*
Effect A—Increases the corner freedoms.
Effect B—Decreases the corner safety locks.
ALTERATION No. 2—*Increasing the drop lock.*
Effect A—Causes the bankings to be spread apart. (Banked to drop.)
Effect B—Increases the corner freedom.
Effect C—Increases the corner safety locks.

RULE 4

311. *Decreasing Corner Freedom.*
ALTERATION No. 1—*Increasing the levers' acting length, or roller jewel radius.*
Effect A—Lessens the corner freedoms.
Effect B—Increases the corner safety locks.
ALTERATION No. 2—*Lessening the drop locks.*
Effect A—The bankings are brought closer together. (Banked to drop.)
Effect B—Lessens the corner freedoms.
Effect C—Lessens corner safety lock.

RULE 5

312. *Increasing Guard Freedom.*—

ALTERATION No. 1—*Increasing the distance separating the guard point from table.*

Effect A—Increases the guard freedoms.

Effect B—Decreases the safety locks.

Effect C—May result in causing either a butting or overbanking error.

ALTERATION No. 2—*Making each drop lock deeper.*

Effect A—The bankings are spread apart. (Banked to drop.)

Effect B—Increases the guard freedoms.

Effect C—Increases the safety locks.

RULE 6

313. *Decreasing Guard Freedom.*—

ALTERATION No. 1—*Lessening the distance between guard point and table.*

Effect A—Decreases guard freedom.

Effect B—Increases the safety locks.

ALTERATION No. 2—*Decreasing the drop locks.*

Effect A—The bankings are brought closer together. (Banked to drop.)

Effect B—Decreases guard freedoms.

Effect C—Decreases the safety locks.

RULE 7

314. *Increasing Levers' Acting Length, or the Roller Jewel Radius.*—

ALTERATION—*Making longer, either the lever's acting length, or roller jewel radius.*

Effect A—Decreases corner freedoms.

Effect B—Increases the corner safety locks.

RULE 8

315. *Shortening Lever's Acting Length, or Roller Jewel Radius.*

ALTERATION—*Decreasing the lever's acting length, or radius of roller jewel.*

Effect A—Increases corner freedom.

Effect B—Decreases the corner safety locks.

RULE 9

316. *Opening the Banking Pins.*—

ALTERATION—*Spreading the bankings apart.*

Effect A—Increases guard freedom.
Effect B—Increases corner freedom.
Effect C—Increases, or provides "slide."
Effect D—Increases, or provides "run."

RULE 10

317. *Closing the Banking Pins.*—
ALTERATION—*Bringing the banking pins closer together.*
Effect A—Lessens guard freedom.
Effect B—Lessens corner freedom.
Effect C—Lessens, or removes "slide."
Effect D—Lessens, or removes "run."
Effect E—Banks to drop. (No run, no slide.)

RULE 11

318. *Banked to Drop.*—An Elgin type of escapement when *banked to drop* will show the following:
(a) Some guard freedom.
(b) Some corner freedom.
(c) A safety lock *less* in amount than the drop lock.

RULE 12

A South Bend escapement, when banked to drop, will show:
(d) *No* guard freedom.
(e) *No* corner freedom.
(f) A safety lock of the *same* amount as the drop lock.

RULE 13

319. *Guard Point Butting Table.*—When it is found impossible to adjust the guard point so as to provide correct guard freedom without introducing a butting error, it compels us to either lessen the diameter of the original table or supply a smaller table.

Either alteration will, when the guard point is advanced, provide correct guard freedom and prevent butting.

RULE 14

320. *Corner Freedom and Drop Lock, Both Defective.*—Should the corner freedoms (banked to drop) be excessive or deficient, and either defect is associated with errors in the drop locks, *first correct the locks.* If still necessary, rectify any error then found in the freedoms, as directed in the Test Lessons and Rules.

RULE 15

321. *Guard Freedom and Drop Lock, Both Defective.*—When the guard freedoms (banked to drop) are either excessive or

deficient, and the defect present is associated with an error in the drop locks, *first correct the locks.* Then, if a guard freedom error still remains, correct it, as instructed in the Test Lessons and Rules.

RULE 16

322. *Protection of the Safety Lock.*—The safety lock is guarded by:

(a) Extent of drop lock.
(b) Amount of guard freedom.
(c) Amount of corner freedom.
(d) In all double roller escapements by central part of lever horn and roller jewel.

NOTE.—The amount of guard freedom, and the amount of corner freedom, also the freedom referred to at D, must in all escapements be *less* than the drop lock.

323. *Escapements Out of Angle.*—Cause of errors: (1) Drop locks unequal, or (2) lever bent.

RULE 17

Angular Test—Any dissimilarity in the position of each tooth with its respective pallet jewel indicates the escapement is out of angle

RULE 18

Corner Test—Should the corner freedoms be *unequal* (banked to drop) and the roller jewel be straight, the escapement is out of angle.

RULE 19

Guard Test—*Any difference* in the guard freedoms expresses the fact that the escapement is out of angle, provided the guard point is straight and escapement banked to drop.

Index to Test Lessons 1 A to 6 D

324. *Corner Test.*—

LIST 1

1 A. Proof findings. Elgin type.

LIST 2

2 A. Drop locks light. Corner freedoms falsely correct.
2 B. Drop locks deep. Corner freedoms falsely correct.

LIST 3

3 A. Drop locks correct. Corner freedoms excessive.
3 B. Drop locks light. Corner freedoms excessive.
3 C. Drop locks deep. Corner freedoms excessive.

LIST 4

4 A. Drop locks correct. Corner freedoms lacking.
4 B. Drop locks light. Corner freedoms lacking.
4 C. Drop locks deep. Corner freedoms lacking.

LIST 5

5 A. Drop locks correct. Roller jewel retained in slot.
5 B. Drop locks light. Roller jewel retained in slot.
5 C. Drop locks deep. Roller jewel retained in slot.

LIST 6

Corner Safety Test.

6 A. Drop locks correct. Corner freedoms correct. Error corner trip.
6 B. Drop locks correct. Corner freedoms excessive. Error corner trip.
6 C. Drop locks light. Corner freedoms correct. Error corner trip.
6 D. Drop locks deep. Corner freedoms excessive. Error corner trip.

LESSON 34

TEST LESSON NO. 1 A. CORRECT OR PROOF FINDINGS—ELGIN TYPE—BANKED TO DROP

325. *Proof-Findings.—Remarks*—An escapement of the Elgin type belonging to the correct or commercially correct class will, if the parts are well matched, show the following conditions:

Drop Lock—The drop or first locks should be equal, light, and safe.

Corner Test—The corner test should show an equal amount of freedom betwixt the roller jewel and each slot corner.

Remarks—The correct amount of corner freedom cannot be learned from a book. The student is advised to test many high grade escapements and thereby discover what is meant by "corner freedom correct." In practice the expression correct corner freedom, like correct lock, is rather elastic; it depends very much on the grade of the watch.

Guard Test—The guard point must have an equal amount of freedom on each side of the table.

Remarks—The term "guard freedom correct," like its fellow term, corner freedom correct, is a variable quantity. Our remarks on correct corner freedoms are equally applicable to correct guard freedoms.

Curve Test—We should by this test find that the roller jewel is free from the tips of the lever horns and can pass into the notch without decidedly catching on any part of the horn or slot corner.

Angular Test—The angular test should show that each tooth barely remains in contact with each pallet jewel, after the manner illustrated in Fig. 29.

Corner Safety Test—Guard Safety Test—Curve Safety Test.—Each of the tests named in our heading, when applied, should show a safety or remaining lock.

LESSON 35

TEST LESSON NO. 2 A—FALSE CORNER FREEDOM—DROP LOCKS LIGHT—ELGIN TYPE—BANKED TO DROP

326. *Corner Test.*—*Remarks*—In a high grade watch, light locks are not productive of escapement trouble, for the reason that all parts are accurately fitted.

In a low grade watch, light locks are a frequent source of trouble. It is to this class of watch that this lesson applies.

Condition of Escapement—Drop locks *light.* Corner freedoms *apparently* correct. Escapement banked to drop.

Observation Test—By observation we discover that the drop locks are unsafely light.

Corner Test—An examination by the corner test reveals an apparently correct amount of corner freedom between the roller jewel and the slot corners.

Remarks—From the facts that the locks are unsafely light, and the corner freedoms apparently correct, we reason that if an increase of the locks *is desirable* altering the locks will cause an *increase* in the corner freedoms.

Alterations—When changes are necessary commence by increasing the drop locks, making same deeper.

Increasing the drop locks means we must spread the bankings, thereby securing a new position of drop lock. *Opening* the bankings *increases* the freedom between the roller jewel and slot corners. (Corner test.)

Remarks—After altering the locks and changing the position of the banking pins the escapement's condition will now read:

Altered Condition of Escapement.—Drop locks commercially correct. Corner freedoms *excessive.* Banked to drop.

Remarks—At times, provided no tripping error develops (corner safety test), further corrections may not be necessary.

Should additional alterations be required to correct an excess of the corner freedom, the necessary changes may be brought about by either supplying a new lever of greater acting length, or by replacing the roller table. The new table must hold the roller jewel in a more forward position. The plan usually followed in the factories is that of changing tables. During the progress of alterations apply the corner, corner safety, and angular tests. When the proof findings of the tests named

r tl matched. See Lesson 34.

LESSON 36

TEST LESSON NO. 2 B—FALSE CORNER FREEDOM—DROP LOCKS DEEP—ELGIN TYPE—BANKED TO DROP

327. *Corner Test.—Condition of Escapement*—Drop locks *deep*. Freedom of roller jewel with slot corners *seemingly* correct. Escapement banked to drop.

Remarks—In this instance an observation of the locks shows them as deep. The corner test when applied indicated an apparently correct amount of corner freedom between the roller jewel and the slot corners. Let us now see the effect of alterations.

Alterations—The Locks—As deep drop lock is a serious defect, we correct it,—making it normal.

The Bankings—Decreasing the drop lock compels us to close in the bankings to a new banked to drop position.

The Corner Test—The corner test now shows that the roller jewel is in contact, or very nearly in contact with the slot corners. This is the result of decreasing the drop locks and the consequent rebanking to drop.

Remarks—Students should learn to reason out effect of changes as indicated in the above alterations. The changes made now show the following results:

Altered Condition of Escapement—Drop locks correct. Freedom *wanting* between roller jewel and slot corners. Escapement banked to drop.

Remarks—The error shown by the above is a lack of corner freedoms. To remedy this either the acting length of the lever must be shortened or the radius of the roller jewel decreased.

Alteration—The corner freedom can be increased by cutting away a part of the horns and slot corners. We can also increase the corner freedom by changing the position of the roller jewel, setting it closer to the center of its table, or by selecting a new table having its roller jewel in the desired position. While changes are being made, frequently use the corner, corner safety, and angular tests to verify correctness of alterations.

LESSON 37

TEST LESSON NO. 3 A—EXCESSIVE CORNER FREEDOM—DROP LOCKS CORRECT—ELGIN TYPE—BANKED TO DROP

328. *Corner Test.—Condition of Escapement*—Drop locks *correct. Excessive* freedom of roller jewel with slot corners. Banked to drop.

Remarks—With an eyeglass we inspect the drop locks and find they are correct. The corner test in this instance shows excessive freedom between the slot corners and the roller jewel.

The drop locks being correct, require no alteration, but we are confronted with this defect—when the escapement is banked to drop the corner test reveals a great deal of space betwen the roller jewel and the slot corners.

Alterations.—The Locks—The drop locks being correct, do not require changing.

Remarks—As the drop locks are satisfactory, we must, to correct the excessive corner freedom, either increase the lever's acting length, or bring the roller jewel more forward, or we can combine both methods.

The Lever and the Roller Jewel—In the better class of watches it is advisable to either advance the position of the roller jewel, or select a new table, one having the roller jewel in a more advanced position. Either operation is easier than the fitting of a new and longer lever. If the watch is of very low grade and possesses a soft lever the walls of the slot can be stretched, thereby increasing the lever's acting length. By so doing, the excess of space found between the slot corners and the roller jewel can be diminished.

Remarks—From a practical standpoint a slight excess of corner freedom need not be regarded as detrimental. Excessive corner freedom must be altered *whenever* the corner safety test decides that the safety or remaining lock is endangered. An examination by the student of a large number of escapements will prove that exactness in corner freedom is frequently lacking. While alterations are in progress apply the corner, corner safety, and angular tests.

LESSON 38

TEST LESSON NO. 3 B—EXCESSIVE CORNER FREEDOM—DROP LOCKS LIGHT—ELGIN TYPE—BANKED TO DROP

329. *Corner Test.*—Condition of Escapement—Drop lock *light. Excessive* freedom between the slot corners and roller jewel. Banked to drop.

Remarks—An inspection of the drop locks in this escapement reveals them as being light. The corner test shows a surplus of freedom between the roller jewel and the slot corners. We have here two defects requiring attention.

Alterations—The Locks—The rule is when the drop locks are defective correct them first. In conformity with this rule we alter the locks, making them deeper and at the same time correct.

The Bankings—Increasing the amount of the drop locks means we draw the pallet jewels further out of their settings. This compels us to open the bankings and establish a new position of drop lock.

The Lever—The act of increasing the locks caused us to spread the bankings more apart. The result is, we have *increased* the previous error of excessive corner freedom. The corner freedoms now being greater than before the locks were altered, the escapement condition is therefore as follows:

Altered Condition of Escapement—Drop locks correct. Corner freedoms have become *more excessive.* Banked to drop.

Remarks—To correct the error of excessive corner freedom compels us to either increase the lever's acting length, or increase the roller jewel radius. Follow the instructions already given and make constant use of the corner, corner safety, and angular tests while changing the escapement parts. (See Lesson 37.)

LESSON 39

TEST LESSON NO. 3 C—EXCESSIVE CORNER FREEDOM—DROP LOCKS DEEP—ELGIN TYPE—BANKED TO DROP

330. *Corner Test.—Condition of Escapement*—Drop locks *deep. Excessive freedom* between the slot corners and the roller jewel. Banked to drop.

Remarks—An observation of the drop locks of this escapement teaches us that the drop locks are rather deep. In addition, we learn from the corner test that a surplus of freedom or space exists under banked to drop conditions between the roller jewel and the slot corners.

Alterations—The Locks—Following the rule the locks are first altered into a more correct form. To do so the pallet stones were pushed further back into their respective settings.

The Bankings—Decreasing the drop locks enabled us to bring the bankings closer together, thereby a new drop lock position is obtained.

The Lever—The results obtained by lessening the locks and the establishment of a closer banked to drop position of the bankings is, the slot corners are brought nearer to the path of the roller jewel and in consequence the excessive corner freedoms are lessened.

Remarks—When a deep drop lock is associated with an excess of corner freedom, as a general rule, changing the deep to a lighter and more correct form of lock, automatically acts as a corrective of surplus corner freedoms.

LESSON 40

TEST LESSON NO. 4 A—CORNER FREEDOM LACKING—DROP LOCKS CORRECT—ELGIN TYPE—BANKED TO DROP

331. *Corner Test.—Condition of Escapement*—Drop locks correct. Corners of the slot and roller jewel either *too close* or in actual *contact*. Banked to drop.

Remarks—The drop locks of this escapement are of a correct type. The corner test (escapement banked to drop) shows either contact or a decrease of the usual amount of corner freedom. The locks being correct, the position of the banking pins will therefore remain unchanged.

The Lever—As no alterations of the locks are necessary or allowable, we must, to correct the want of corner freedoms, either shorten the lever's acting length or in some manner decrease the radius of the roller jewel.

Remarks—As a precaution against the introduction of other errors, freely use the corner, corner safety and angular tests. As before mentioned, the corner test shows an absence or lack of the usual amount of corner freedom. As the drop locks are *correct* the angular test shows that each tooth is discharged from its respective pallet jewel. When we find the drop locks correct and corner freedom lacking we can assign the cause to either of the following: (a) the lever's acting length is too long or (b) the radius of the roller jewel is too great. Shortening the roller jewel radius is the plan usually followed in the factories. (See Lesson 36.)

LESSON 41

TEST LESSON NO. 4 B—CORNER FREEDOM LACKING—DROP LOCKS LIGHT—ELGIN TYPE—BANKED TO DROP

332. *Corner Test.—Condition of Escapement*—Drop locks *light.* The corners of the slot and the roller jewel are either *very close* or come *in contact.* Banked to drop.

Remarks—By observation we determine that the drop locks are light and unsafe. The corner test discloses a shortage of freedom—that is, the roller jewel is in close proximity to the slot corners.

Alterations—The Locks—As there is an error in the drop locks, we give it first attention. Accordingly we increase the drop locks.

The Bankings—Increasing the drop locks compel us to spread the bankings apart.

The Lever—The effect of opening the bankings to a new position of drop lock, provides more freedom between the roller jewel and slot corners. To confirm the changes use the corner and angular tests.

LESSON 42

TEST LESSON NO. 4 C—CORNER FREEDOM LACKING—DROP LOCKS DEEP—ELGIN TYPE—BANKED TO DROP

333. *Corner Test.—Condition of Escapement*—Drop locks *deep.* Slot corner and roller jewel *in contact* or nearly so. Banked to drop.

Remarks—As above stated, the drop-locks are too deep. The corner test, the watch being banked to drop, brings out the fact that but little if any freedom can be found between the slot corners and the roller jewel.

Alteration—The Locks—Our first alteration is the locks; these we decrease and make correct.

The Bankings—To maintain banked to drop positions we must, when the drop locks are lessened, turn in each banking pin

The Lever—In this escapement, lessening the locks and re-banking to drop makes matters worse, the corner freedom being still further decreased. To remedy the want of corner freedom, we must either shorten the lever by cutting away the horns and slot corners, or else try shifting the roller jewel closer to the center of its table. Check the changes by the corner and angular test. When conditions equivalent to "proof-findings" of the angular test are obtained we realize that the length of the lever is adapted to the extent of the drop lock.

LESSON 43

TEST LESSON NO. 5 A — ROLLER JEWEL RETAINED IN SLOT—DROP LOCKS CORRECT —BANKED TO DROP—ELGIN TYPE

334. *Condition of Escapement.*—Drop locks correct. Roller jewel *unable* to leave slot. Banked to drop.

Remarks—Observation shows the drop locks are correct. With the escapement *banked to drop* the roller jewel we find is *unable* to make its exit out of the slot. Given the above conditions, we reason as follows:

The drop locks are correct, therefore the present banked to drop position of the banking pins *cannot* be changed. The roller jewel, however, is unable to escape out of the slot.

Alterations—Under the existing escapement conditions—namely, locks correct and the roller jewel held by the slot—we are led to decide that either the lever's acting length is too long or the roller jewel is set too far forward, or combinedly they cause the error. It therefore can be reasoned out that if we cut away a part of the horns and slot corners the roller jewel will be able to emerge out of the slot. Don't neglect constant use of the corner and angular tests to decide when normal conditions have been met.

LESSON 44

TEST LESSON NO. 5 B — ROLLER JEWEL RETAINED IN SLOT—DROP LOCKS LIGHT—BANKED TO DROP—ELGIN TYPE

335. *Condition of Escapement*—Drop locks *light.* Roller jewel *unable* to leave slot. Banked to drop.

Remarks—The drop locks we learn are unsafely light. When the escapement is banked to drop the roller jewel is held in the slot and unable to make its slot.

Alterations—Alterations are commenced by making the locks deeper. Increasing the locks compels us to rebank to drop. In this instance rebanking means, first, the spreading of the bankings apart; second, an increase of the corner freedom. These alterations *may* provide the correct amount of corner freedom. In the event of the corner freedom still being deficient, either the lever's length will have to be shortened, or the roller jewel set nearer the center of the table. The corner and angular test will decide when the escapement is in good condition.

LESSON 45

TEST LESSON NO. 5 C — ROLLER JEWEL RETAINED IN SLOT—DROP LOCKS DEEP—BANKED TO DROP—ELGIN TYPE

336. *Condition of Escapement.*—Drop locks *deep.* Roller jewel *unable* to make exit out of slot. Banked to drop.

Remarks—By observation we decide that the drop locks are deep. When the escapement is banked to drop we find that the roller jewel is unable to get out of the slot.

Alterations—The drop locks require first attention; on making the locks lighter we are thereby compelled to rebank to drop. Rebanking in this instance brings the banking pins *closer* together, and it further decreased the chance of the roller jewel to emerge out of the slot. Therefore to obtain the release of the roller jewel, so that it can take its part in the escapement action, the lever's length must be made shorter, or the radius of the roller jewel decreased as formerly described. The corner and angular tests should be used to confirm changes.

LESSON 46

TEST LESSON NO. 6 A — CORNER TRIPPING ERROR—ELGIN TYPE

337. *Corner Safety Test.—Condition of Escapement*—Banked to drop. Drop locks correct. Corner test shows corner freedoms correct. Corner safety test develops a *tripping error*.

Remarks—The Locks—The drop locks in this escapement are correct.

Corner Test—The freedom of the roller jewel with the slot corners (corner freedom) is satisfactory.

Corner Safety Test—By means of the corner safety test we find that the *length* of the escape wheel teeth *vary*, because some of the teeth show a safety lock, while others leave the pallet jewel's locking face. This latter condition being a corner trip, is of course an error.

Alterations—The remedy, if the watch is worthy of it, is a new escape wheel, one possessing teeth regular in length. If the watch is a poor one, or should the charges not warrant a new wheel, the tripping can be overcome by increasing the locks. As a check on the result of alterations use the corner, corner safety, and angular tests.

LESSON 47

TEST LESSON NO. 6 B — CORNER TRIPPING ERROR—ELGIN TYPE

338. *Corner Safety Test.—Condition of Escapement*—Banked to drop. Drop locks correct. Corner freedoms *excessive*.. A corner *trip present*.

Remarks—The locks, as an inspection shows, are practically correct. The corner test reveals that *too much* freedom exists between the slot corners and the roller jewel. When the corner safety test is applied a corner trip is discovered.

Alterations—If the corner freedom is only a trifle excessive and the resultant corner trip is also very slight, the corner trip can be corrected by either of the following methods: (a) slightly increasing the locks; (b) slightly advancing the position of the roller jewel.

If the corner trip is of a most decided character, the drop locks *being correct*, we would then be compelled to either increase the lever's acting length, or bring the roller jewel more forward. By using one, or both methods combined, the corner trip can be eliminated. The corner, corner safety, and angular tests should be constantly employed to check whatever changes are made.

LESSON 48

TEST LESSON NO. 6 C — CORNER TRIPPING ERROR—ELGIN TYPE

339. *Corner Safety Test.—Condition of Escapement*—Banked to drop. Drop locks *light*. Corner freedoms correct. *Corner trip present.*

Remarks—When we discover a corner trip, the drop locks being light, but with the correct amount of corner freedoms present, we must, to correct the tripping error, increase the drop locks.

Alterations—Making the drop locks deeper causes us to open the bankings. The result of these changes is to slightly increase the corner freedoms.

If after increasing the locks and rebanking the escapement to drop we find that the new corner freedoms are not too excessive, the surplus in the freedoms may be ignored. Unless proven to be detrimental by the corner, corner safety, and angular tests, in which event it requires to be remedied as directed in preceding lessons.

LESSON 49

TEST LESSON NO. 6 D — CORNER TRIPPING ERROR—ELGIN TYPE

340. *Corner Safety Test.—Condition of Escapement*—Banked to drop. Drop locks *deep*. Corner freedoms *excessive*. Corner trip present.

Remarks—A complication of errors exist in this escapement. First, the drop locks are deep; second, there is too much freedom of the roller jewel with each slot corner; third, a corner trip is present.

Our first alteration is that of decreasing the drop locks, changing them into a more correct form. The effect of lessening the locks is as follows: (a) There is a slight decrease in the excessive corner freedoms; (b) the tripping error is increased rather than decreased.

We have now made the locks correct, but further changes are necessary to decrease both the excessive corner freedoms and the tendency of the parts to produce a trip. The required alterations consist in making the lever's acting length longer or advancing the position of the roller jewel, as previously described. To confirm correctness of changes, use the corner, corner safety, and angular tests.

Index to Test Lessons 7 A to 12 A

341 A. *Guard Test and Guard Safety Test.*

LIST 7

7 A. Drop locks light. Guard freedoms apparently correct.
7 B. Drop locks deep. Guard freedoms apparently correct.

LIST 8

8 A. Drop locks correct. Guard freedoms excessive.
8 B. Drop locks light. Guard freedoms excessive.
8 C. Drop locks light. Guard freedoms excessive.

LIST 9

9 A. Drop locks correct. Guard freedoms lacking.
9 B. Drop locks light. Guard freedoms lacking.
9 C. Drop locks light. Guard freedoms lacking.

LIST 10

10. Guard point butts table.

LIST 11

11 A. Drop locks correct. Guard freedoms correct. Guard trip error.
11 B. Drop locks correct. Guard freedoms excessive. Guard trip error.
11 C. Drop locks light. Guard freedoms correct. Guard trip error.
11 D. Drop locks light. Guard freedoms excessive. Guard trip error.

LIST 12

12 A. Drop locks deep. Guard freedoms excessive. Guard trip error.

LESSON 50

TEST LESSON NO. 7 A—FALSE GUARD FREEDOM

341 B. *Condition of Escapement.*—Drop locks light. Guard freedoms *apparently* correct. Banked to drop.

The Locks—We assume that in this escapement, the drop locks are unsafely light.

The Guard Freedoms—The escapement being banked to drop, we find, by means of the guard test, that the freedoms between the guard point and edge of table are apparently correct.

Alterations—As we judged the drop locks to be too light, we increase them. Making the drop locks deeper, compels us to open out each banking—that is, we rebank the escapement to drop. The effect of spreading the banking pins is to *increase* the guard freedoms.

The effect of the alterations is to place the escapement in the following condition:

Altered Condition—Drop locks correct. Guard freedoms *excessive*. Banked to drop.

The locks are now correct; but the guard freedoms are greater than before alterations were made. To remedy the increase in the guard freedoms, the guard point should be advanced closer to the table, or else obtain a new table of slightly greater diameter.

Remarks—Whenever the guard freedom is altered, as a check, compare it with the corner freedom. Theoretically these freedoms should be equal, and from a practical standpoint it is wise, although not always practicable or possible, to maintain them so. If any difference is favored the corner freedom should be a trifle the greatest.

In double roller escapements, when changes affecting the guard point are made, the curve test should be consulted for reasons explained elsewhere. (Compared with Lesson 35.)

LESSON 51

TEST LESSON NO. 7 B—FALSE GUARD FREEDOM

342. *Guard Test.*—Elgin type. Banked to drop.

Condition of Escapement—Drop locks *deep*. Guard freedoms *apparently* correct. Banked to drop.

The Locks—We discover by observation that the drop locks are deep.

The Guard Freedoms—With the escapement banked to drop, the freedoms between the guard point and table appear satisfactory.

Alterations—The deep drop locks require first attention, accordingly we change them into a more correct form of lock. Lessening the drop locks necessitates the closing of the bankings.

In this instance the effect of rebanking to drop is to lessen the freedom of the guard point with the table.

Assuming that the changes made bring about contact, or near contact, of the guard point with the table, the escapement's condition will appear as follows:

Altered Condition—Drop locks correct. Guard freedoms *none*. Banked to drop.

To correct the lack of guard freedom, the guard point must be set further back—that is, removed from the table; or a new table lesser in diameter should be supplied.

At times it is permissible to use the lathe for turning away the edge of the original table. In this manner the diameter of the table can be lessened. Care should be taken to highly repolish its edge.

Remarks—If in an effort to provide the requisite guard freedoms, the escapement being banked to drop, we bend the guard point away from the edge of the table, and find when the guard test is tried, that the guard point is inclined to stick or jam against the edge of the table, our best procedure then is to readvance the guard point to its former position; then, to provide the necessary freedom, the old table must be lessened in diameter, or a new table used, whose diameter provides the correct amount of guard freedom. Butting errors if developed always require correction. (Compare with Lesson 36.)

LESSON 52

TEST LESSON NO. 8 A—EXCESSIVE GUARD FREEDOM

343. *Guard Test.*—Elgin type. Banked to drop.

Condition of Escapement—Drop locks correct. Guard freedoms *excessive.* Banked to drop.

The Locks—As above stated, the drop locks in this escapement are correct.

Guard Test—The guard test shows there is too much freedom between the guard point and table.

Alterations—As the drop locks are classed as correct, we pass on to the error of excessive guard freedom. To correct this error, we must either advance the guard point towards the table, or supply a new table of greater diameter. Whichever course we follow will remedy the error.

As previously directed, compare the corner and guard freedoms—make use of the guard safety test, and in addition, if the escapement is of the double roller type, the curve test should be employed.

LESSON 53

TEST LESSON NO. 8 B—EXCESSIVE GUARD FREEDOM

344. *Guard Test.*—Elgin type. Banked to drop.

Condition of Escapement—Drop locks *light.* Guard freedom *excessive.* Banked to drop.

The Locks—The first fault we encounter is that of light drop locks.

The Guard Freedoms.—The second fault, as uncovered by the guard test, is, there exists too much play or freedom between the guard point and edge of table.

Alterations—The first items calling for correction are the drop locks. These we increase, making them deeper. Increasing the locks requires us to spread the banking pins more apart, and by doing so a new banked to drop position is established.

The act of spreading the bankings increases the freedom between the guard point and table. This means we have *increased* the error of excessive guard freedom.

To correct excessive guard freedom the guard point must be brought closer to the edge of the table, or by supplying a new table greater in diameter; this will also provide the correct amount of guard freedom. (Compare with Lesson 38.)

LESSON 54

TEST LESSON NO. 8 C—EXCESSIVE GUARD FREEDOM

345. *Guard Test.*—Elgin type. Banked to drop.

Condition of Escapement—Drop locks *deep.* Guard freedom *excessive.* Banked to drop.

The Locks—By observation, and by banking the escapement to drop, we decide that the drop locks are too deep.

The Guard Freedom—The guard test informs us that there is too much freedom between the guard point and the table.

Alterations—Directing our attention to the deep locks, we reduce them to the standard of correct lock. Changing the locks allows us to close in each banking in accordance with banked to drop rules. By closing in the banking the excessive guard freedom is lessened, and whether now correct or not depends on actual conditions. (Compare with Lesson 39.)

LESSON 55

TEST LESSON NO. 9 A—GUARD FREEDOM WANTING

346. *Guard Test.*—Elgin type. Banked to drop.

Condition of Escapement—Drop locks correct. Guard freedom *none* or very little. Banked to drop.

The Locks—In this escapement the drop locks are correct. We use the term "correct" to express the idea that the locks are "commercially right" and therefore satisfactory.

The Guard Freedoms—The guard point possesses little if any freedom with the edge of the table.

Alterations—The drop locks being classed as correct, require no changing.

When the guard test was applied we therefrom learned that the customary amount of guard freedom was lacking.

To overcome this state of contact, or of near contract, we must either *remove* the guard point from the table, or supply a *new table* of lesser diameter.

The guard safety test, and the Rules preceding the Test Lessons, should be consulted, if difficulty is experienced about obtaining a correct adjustment of the guard point to the table. (Compare with Lesson 40.)

LESSON 56

TEST LESSON NO. 9 B—GUARD FREEDOM WANTING

347. *Guard Test.*—Elgin type. Banked to drop.

Condition of Escapement—Drop locks light. Guard freedom *none* or very slight. Banked to drop.

The Locks—We discover the drop locks are light.

The Guard Freedoms—The guard test shows that guard freedom is lacking.

Alterations—Our first move is to alter the locks, making same deeper, and at the same time correct. Increasing the drop locks compel us to spread the bankings to a new drop lock position. This change in the location of the banking pins, incidentally provides freedom between the guard point and the table. When the guard freedom approaches the extent of the corner freedom, it will be in agreement with correct escapement conditions. Should a butting error develop, Lesson No. 58 will be found helpful. (Compare with Lesson 41.)

LESSON 57

TEST LESSON NO. 9 C—GUARD FREEDOM WANTING

348. *Guard Test.*—Elgin type. Banked to drop.

Condition of Escapement—Drop locks *deep*. Guard freedom *little*, if any. Banked to drop.

The Locks—An inspection of the drop locks, under banked to drop conditions shows they are deep.

Guard Freedoms—By the guard test we learn that but little if any freedom exists between the guard point and the table.

Alterations—We first change the deep to a correct drop lock. Decreasing the locks call for a rebanking to drop, the effect of which is to bring the banking pins closer together. The result of turning in the bankings is to make matters worse, the guard freedom being further lessened.

To correct the want of guard freedom we must either remove the guard point away from the table or replace the old table with one lesser in diameter. Another remedy is, place the old table in the lathe and reduce its circumference. (Compare with Lesson 42.)

LESSON 58

TEST LESSON No. 10—BUTTING ERROR—ELGIN TYPE

349. *Guard Point Jams or Butts Against Edge of Table.*—When, on making the guard or similar test, a workman finds that the guard point remains, or is inclined to remain in contact with the edge of the table, it is an indication of a butting error.

Butting errors may be divided into three classes, as follows: First, errors due to defective draw; second, errors caused by the pivots of pallet or balance staff working in holes too large for them; third, a butting error arising from a guard point defectively placed, or caused by some fault in the *size* of the table.

Butting errors are readily discovered without the escapement being banked to drop. Once this error is found it is of the utmost importance that the escapement be banked to drop; by doing so the exact cause of the trouble can be located and corrections made. Butting errors must never be allowed to go uncorrected, or stoppage will ensue.

Example 1—We shall assume an instance wherein after banking to drop, that too much guard freedom is found combined with a butting error. There is a likelihood that by advancing the guard point closer to the table, that the butting trouble can be overcome.

It must be understood that a limit exists to the extent we can infringe on the guard freedom. While bringing the guard point closer to the table lessens the error, it may also injuriously affect the escapement, owing to the normal amount of guard freedom being decreased.

Example 2—When a butting error can only be overcome by placing the guard point unduly close to the edge of the table, we must, to obtain the correct amount of guard freedom, either lessen the size of the table, or provide a new table of lesser diameter. The plans suggested will supply the necessary amount of guard freedom, and abolish the butting tendency.

Example 3—Some escapements whose drop locks are correct, aud having an apparently correct amount of guard freedom, show an inclination of the guard point to butt against the edge of the table. To correct the difficulty the table's diameter must be made less, and the guard point advanced toward the table's edge. Both alterations are necessary to correct the butting, and secure the requisite guard freedoms.

In many escapements, butting errors are easily created, simply by bending the guard point a slight amount away from edge of table. Students are therefore advised to thus start a practical investigation of such errors.

We wish to call attention to a type of butting error frequently met with—namely, (a) in setting a watch the seconds hand is sometimes pushed backwards, the result often being that the watch stops; or, (b) when placing the seconds hand on post of fourth wheel stoppage follows. When such happens it is usually attributable to the guard point for some reason "butting" or wedging against edge of table. The exact cause of error should be located and correction made.

LESSON 59

TEST LESSON NO. 11 A—GUARD TRIPPING ERROR

350. *Guard Safety Test.*—Elgin type.

Condition of Escapement.—Drop locks correct. Guard test shows correct guard freedom. Guard safety test develops *a trip.*

The Locks—The drop locks in this escapement are correct.

The Guard Test—We learn from the guard test that a correct amount of guard freedom is present.

The Guard Safety Test.—The guard safety test shows a tripping error, involving only some teeth of the escape wheel.

Alterations—As a guard tripping error is present and as some of the teeth show a trip while others show a safety lock, it is clear that the cause is due to the teeth of the escape wheel being irregular in length. To remedy the trouble, supply a new escape wheel having teeth regular in length. Should the price obtain for repairing not be sufficient to cover the expense of a new wheel, we can continue the use of the old wheel by increasing the drop locks just enough to prevent tripping.

Remarks—It is always advisable to examine the guard point; if loose, make it tight, and thereby avoid possible escapement troubles.

The following advice will also be found profitable: Whenever, as a matter of testing, we bring the guard point in contact with one side of the table and a tripping error is found, and further tests on this particular side of the table agree in exposing more trips; while like tests on the opposite side of the table, all show safety locks, it would lead us to suspect that for some reason the edge of the table is running out of truth, or the guard point is bent to one side, or if straight, it is defective in shape. Of course faulty parts must be changed and retests made. (Compare with Lesson 46.)

LESSON 60

TEST LESSON NO. 11 B—GUARD TRIPPING ERROR

351. *Guard Safety Test.*—Elgin Type.

Conditions of Escapement—Drop locks correct. Guard freedom *excessive.* Guard safety test exposes *a trip.*

The Locks—As stated above, the drop locks are correct.

Guard Test—The guard test shows that the freedom between guard point and table is excessive.

The Guard Safety Test—The guard safety test shows that the teeth of the escape wheel do not remain on the pallet stone's locking face, but enter slightly on to the pallet's impulse face, thereby denoting that a tripping error is present.

Alterations—The excessive guard freedom is the cause of the trip. If the extra amount of guard freedom is attributable to a guard point being inclined away from the table, then to correct the tripping error, and overcome the excessive guard freedom, the guard point must be straightened.

If we find that the guard point is straight, the locks being correct, with a trip present, the instructions given in Test Lesson No. 11 A will apply. (Compare with Lesson 47.)

LESSON 61

TEST LESSON NO. 11 C—GUARD TRIPPING ERROR

352. *Guard Safety Test.*—Elgin Type.

Condition of Escapement—Drop lock *light.* Guard freedoms correct. Guard *tripping error* present. Banked to drop.

Remarks—If a guard tripping error is discovered, when the

drop locks are light, and the freedom between the guard point and table is satisfactory, we can, by increasing the locks, overcome the tripping error.

Alterations—The locks being light, we increase them. The effect of making the drop locks deeper brings about two results—first, the tripping fault is overcome; second, rebanking to drop compels us to spread the banking pins more apart. The effect of spreading the bankings is to increase the former correct freedoms into excessive freedoms. The usual experience is, that a slight increase in the guard and corner freedoms incident to a slight increase of the drop locks is not detrimental. Should the freedoms be injuriously excessive, consult Test Lessons 3 A and 8 A. (Compare with Lesson 48.)

LESSON 62

TEST LESSON NO. 11 D—GUARD TRIPPING ERROR

353. *Guard Safety Test.*—Elgin type.

Condition of Escapement—Drop locks *light.* Guard freedoms *excessive.* Guard *tripping error* present. Banked to drop.

Remarks—When we discover a guard trip, combined with too much freedom between guard point and table, the drop locks being too light, the following changes are necessary:

Alterations—We first increase the drop locks; this, when we rebank to drop, causes us to spread the bankings. The escapement will then be in the following condition:

Altered Condition—Drop locks correct. Guard freedom *excessive.* Banked to drop.

To correct the excessive guard freedom consult Test Lessons Nos. 8 A-8 B. Should the corner freedoms be likewise excessive see Test Lessons No. 38.

LESSON 63

TEST LESSON NO. 12 A—GUARD TRIPPING ERROR

354. *Guard Safety Test.*—Elgin type.

Condition of Escapement—Drop locks deep. Guard freedoms *excessive.* Guard *tripping error* present. Banked to drop.

Remarks—It is possible, even with deep drop locks, to discover guard tripping errors, when there exists too much freedom between the guard point and the table.

Alterations—The first change required is that of altering the deep into a correct form of lock. This done, the condition of the escapement will be:

Altered Condition—Drop locks correct. Guard freedoms still *excessive.* Guard *tripping error* present. Banked to drop.

Remarks—It is very likely that errors of excessive guard freedom, and guard trips, will be found associated with errors of excessive corner freedom and corner trips. The corrections indicated are outlined in Test Lessons Nos. 8 A, 3 A and 6 B. (Compare Lesson 49.)

Index to Test Lessons 13 A to 18 A

355. *Angular Test.*—

LIST 13

LESSON 64

TEST LESSON NO. 13 A—ELGIN TYPE STATEMENT OF FINDINGS BY THE ANGULAR TEST WITH INDICATED CORRECTIONS

Remarks—*Before* using the angular test *correct* any defects found in the drop locks. Follow the instructions about first *correcting* the locks and you will be able with the assistance of statements Nos. 1, 2 and 3 to determine the nature of the fault. When the locks are corrected, faults if found, are then attributable to some defect in either the lever's acting length, *or* the radius of the roller jewel, or both.

When the locks are *not* corrected beginners are apt to become confused over the test findings. Students should also study Statements 4, 5 and 6. The latter—viz., Statement No. 6—is a very exacting method for determining whether an escapement is out of angle.

WHEN THE LOCKS ARE CORRECT

356. *Angular Test*—Elgin type.

Statement No. 1—Drop locks correct. Lever's acting length correct (matches the lock). Result as shown by test. Tooth and pallet show contact (proof-findings) as indicated by Fig. 29.

Statement No. 2—Drop locks *correct*. Lever's length *long* (or roller jewel's radius long). Result shown by test. Each tooth discharged from pallet. (See Fig. 30.)

Remedies—To establish proof-findings, shorten the lever's length, or radius of roller jewel, or both.

Statement No. 3—Drop locks *correct*. Lever's length *short* (or roller jewel's radius short). Result shown by test. Each tooth shows too much contact with its pallet jewel. (See Fig. 31.)

Remedies—To establish proof-findings, increase lever's length, or roller jewel radius, or both.

WHEN THE LOCKS ARE INCORRECT

Angular Test—Elgin type.

Statement No. 4—Drop locks *light*. Lever's length correct.

Result shown by test. Each tooth discharged from pallet. (See Fig. 30.)

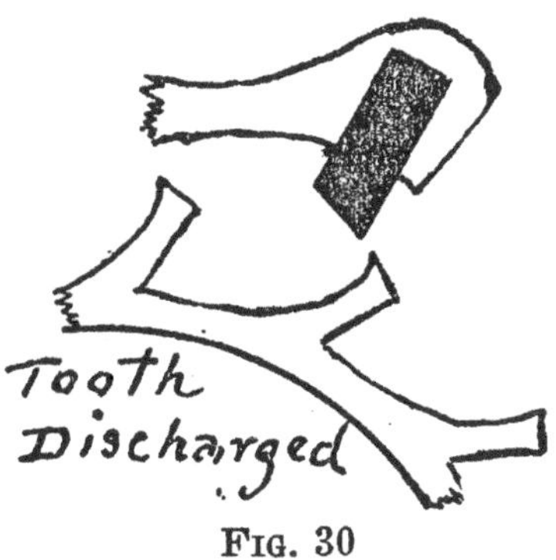

FIG. 30

Remedy—To establish proof-findings increase the drop locks.
Statement No. 5—Drop locks *deep*. Lever's length correct.

FIG. 31

Result shown by test. Each tooth has too much contact with its pallet. (See Fig. 31.)

Remedy—To establish proof-findings lessen the drop locks.

OUT OF ANGLE FINDINGS

Statement No. 6—Drop locks *unequal*. Results obtained by test: (1) On one pallet, tooth shows too much contact (see Fig. 31); (2) tooth is discharged from the opposite pallet (see Fig. 30).

Or,

The teeth occupy dissimilar positions on their respective pallet jewels. *For* instance, one tooth might remain in contact near the letting off corner of one pallet, while the other tooth on opposite pallet might be found near the pallet's entering corner.

Remedy—Equalize the drop lock, then retest.

LESSON 65

TEST LESSON NO. 14 A—PROOF FINDINGS—ELGIN TYPE

357. *Angular Test.*—When by the angular test we discover that each tooth remains in contact with its pallet jewel, at a point very close to the releasing corner of the jewel (see Fig. 29), we feel sure the parts are well matched.

If in addition to the proof findings, as expressed by the slight contact of each tooth with its pallet stone, we find that the drop lock is as light as is consistent with the construction of the escapement—that is, the drop lock belongs in either the correct or commercially correct class—we then know that the parts associated with the angular test are reasonably perfect and that their action will be satisfactory.

LESSON 66

TEST LESSON NO. 15 A—ERROR, DROP LOCKS LIGHT

358. *Angular Test.*—Elgin type.

Condition of Escapement—Drop locks *light.* Lever's length correct.

When the drop locks are light and the lever's acting length is correct, the angular test will exhibit the error by discharging each tooth from each pallet. (See Fig. 30.)

If we increase the drop locks, making them normal, we will find, on repeating the angular test, that each tooth remains in slight contact with its pallet. (See Fig. 29.) The contact taking place near the letting-off corners of each pallet stone, in conformity with the proof findings.

LESSON 67

TEST LESSON NO. 16 A—ERROR, DEEP DROP LOCKS

359. *Angular Test.*—Elgin type.

Condition of Escapement—Drop locks *deep.* Lever's length correct.

Given the drop locks as deep, and the lever's acting length as correct, the angular test will show a surplus of contact of the teeth with their respective pallet jewels. (See Fig. 31.) The *position* occupied by the teeth on each pallet stone's impulse face, depends altogether upon how deep is the drop lock.

By reducing the deep locks into a more correct form we can secure the proof findings of the angular test. (See Statement No. 5, Test Lesson No. 13 A.)

LESSON 68

TEST LESSON NO. 17 A—ERROR, ACTING LENGTH OF LEVER LONG

360. *Angular Test.*—Elgin type.

Condition of Escapement—Drop locks correct. Lever's length *long.*

If the drop locks are correct and the lever's acting length is long, the angular test will show the teeth as discharged from their respective pallet jewel's. (See Fig. 30.)

The remedy is to shorten the lever. During the progress of alterations, make frequent use of the angular test, thereby avoid cutting the lever too short. (See Statement No. 2, Test Lesson No. 13 A.)

LESSON 69

TEST LESSON NO. 18 A—ERROR, ACTING LENGTH OF LEVER SHORT

361. *Angular Test.*—Elgin type.

Condition of Escapement—Drop locks correct. Lever's length *short.*

When the drop locks are correct and the lever's acting length is short, the angular test will show over contact of the teeth with their respective pallet jewels (see Fig. 31)—that is, contact takes place at a point some distance from the letting off corner of each pallet stone.

The remedy required is a new lever of greater acting length, or try advancing the roller jewel more forward; consult the following: Statement No. 3, Test Lesson No. 13 *A*; also Test Lesson No. 3 A.

Index to Test Lessons 19 A to 21 C

Out of Angle

362. *Out of Angle.*—

LIST 19

19 A. Out of Angle Test Findings.
19 B. Drop Locks Unequal. Lever Straight.
19 C. Drop Locks Unequal. Lever Bent.

LIST 20

20 A. Drop Locks Unequal. Lever Straight. Guard Freedoms Unequal.
20 B. Drop Locks Unequal. Lever Bent. Guard Freedoms Unequal.

LIST 21

21 A. Drop Locks Unequal. Lever Straight. Corner Freedoms Unequal.
21 B. Drop Locks Unequal. Lever Bent. Corner Freedoms Unequal.

LESSON 70

TEST LESSON NO. 19 A—ELGIN TYPE

363. *Out of Angle.—An escapement is out of angle:*
(a) When the drop locks are unequal.
(b) When the guard freedoms are unequal.
(c) When the corner freedoms are unequal.
(d) When the angular test shows *dissimilar* positions of each tooth with its respective pallet jewel.

LESSON 71

TEST LESSON NO. 19 B—ERROR, OUT OF ANGLE

364. *Angular Test.*—Elgin type.

Condition of Escapement—Drop locks unequal. Lever straight.

When the drop locks are found to be unequal—viz., the drop lock on one pallet stone exceeds the drop lock on the opposite stone—we immediately know that the escapement is out of angle. As the lever quoted above is straight, the defect is wholly attributable to the faulty locks.

By using the angular test to expose the out of angle condition of the escapement, we will find that the teeth occupy dissimilar positions on their respective pallet jewels.

As the lever is straight, the correction required is that of equalizing the drop locks. This alteration will place the escapement in angle, and cause it to conform with the proof findings expressed in Test Lesson No. 14 A.

LESSON 72

TEST LESSON NO. 19 C—ERROR, OUT OF ANGLE

365. *Angular Test*—Elgin Type.
Condition of Escapement—Drop locks *unequal.* Lever *bent.*

In this escapement not only are the drop locks unequal, but the lever is bent.

The first altertation is that of straightening the lever. This *may* have the effect of equalizing the drop locks. Should straightening the lever *fail* to equalize the locks, we must, in order to place the escapement in angle, reset the pallet jewels. The pallet stones should be reset to meet the requirements of correct lock.

LESSON 73

TEST LESSON NO. 20 A—ERROR, OUT OF ANGLE

366. *Guard Test.*—Elgin type. Banked to drop.
Condition of Escapement—Drop locks *unequal.* Lever straight. Guard point straight. Guard freedoms unequal. Banked to drop.

When the lever and guard point are straight, and the drop locks and guard freedoms are unequal we realize that the escapement is out of angle.

Given an escapement in the above condition, we can, by making the drop locks equal, correct the unequal guard freedoms; this change places the escapement in angle.

As an additional confirmative that the escapement is in angle make use of the angular test.

LESSON 74

TEST LESSON NO. 20 B—ERROR, OUT OF ANGLE

367. *Guard Test.*—Elgin type. Banked to drop.

Condition of Escapement—Drop locks *unequal.* Lever bent. Guard point straight. Guard freedoms unequal. Banked to drop.

The drop locks in this escapement are unequal, and the lever is bent. The guard test, with the escapement banked to drop shows a greater amount of freedom on one side than on opposite side of table. Both appearance and test show that the escapement is out of angle.

In all cases, where we find that the lever is bent, the first alteration is that of straightening the lever; this change is *always* followed by a rebanking of the escapement to drop, and a reinspection of the drop locks. If after straightening the lever the drop locks still remain unequal or otherwise imperfect the pallet stones must be reset in conformity with the requirements of correct lock—viz., a drop lock suited to the particular escapement in hand. When satisfactory alterations have been made the guard freedoms, like the drop locks, will be equalized.

Whenever we find that a lever is bent only a fractional amount, and the resulting inequalities of the drop locks are very slight, corrections are seldom necessary.

When the lever is bent to such an extent that the bend is plainly visible, it is advisable to straighten it, besides making other corrections if required.

At times we find an escapement with the lever bent, but having equal drop locks; when this occurs, conditions, and tests will decide what changes will improve the escapement.

LESSON 75

TEST LESSON NO. 21 A—ERROR, OUT OF ANGLE

368. *Corner Test.*—Elgin type. Banked to drop.

Condition of Escapement—Drop locks unequal. Lever straigh.t Corner freedoms unequal. Roller jewel straight. Banked to drop.

We have in this instance, to contend with unequal drop locks and unequal corner freedoms. This inequality of the freedoms means that the escapement is out of angle. As the lever is straight, the cause of the escapement being out of angle is due to the drop locks being unequal; by correcting the error in the drop locks the corner freedoms will be equalized. Altering the locks therefore places the escapement in angle; this can be confirmed by the angular and guard test.

LESSON 76

TEST LESSON NO. 21 B—ERROR, OUT OF ANGLE

369. *Corner Test.*—Elgin type. Banked to drop.

Condition of Escapement—Drop locks *unequal.* Lever bent. Roller jewel straight. Corner freedoms *unequal.* Banked to drop.

The conditions state three errors—viz., unequal drop locks, unequal corner freedoms, and the lever is bent. These faults we must rectify to overcome the error of out of angle.

The first alterations are, straightening the lever, and rebanking to drop. Should the locks remain unequal after making the lever straight it would be necessary to make the drop locks equal and otherwise correct.

Assuming we have made the corrections, we will find by the corner test that the corner freedoms are equal. The guard test will show the guard freedoms are also equal, and the angular test will further confirm the verity of our work.

Index to Test Lessons 22 A to 26 A

CURVE TEST AND CURVE SAFETY TESTS

370. *Curve Test and Curve Safety Tests.*—

LIST 22

22 *A*. End of Horn. Double Roller. Curve Test.
22 B. Central Part of Horn. Double Roller. Curve Test.
22 C. Slot Corners. Double Roller. Curve Test.
23 *A*. Curve Safety Test. Double Roller.
24 *A*. End of Horn. Single Roller. Curve Test.
25 *A*. Horn and Slot Corners. Single Roller. Curve Test.
26 *A*. Curve Safety Test. Single Roller.

LESSON 77

TEST LESSON NO. 22 A—END OF HORN—DOUBLE ROLLER

371. *Curve Test.*—

NOTE—The following lessons on the curve and curve safety tests, as applied to single and double roller escapements, are a repetition of former explanations. The facts which the lessons contain must be known to you.

To *learn if the roller jewel can touch the extreme end of horn:*

Method—Revolve the balance so as to bring the roller jewel beyond the tip of the horn, then hold the balance secure, next with a fine broach or watch oiler lift the lever off its bank, thereby causing contact of the guard point with edge of safety table. *Maintain the parts in contact* while you slowly rotate the balance so as to bring the roller jewel back past the end of the horn, and towards the lever slot. If these parts are correctly related, the roller jewel as it passes will be *free* from the *end* of the horn. While a rub of the parts when passing is allowable, anything resembling a catch calls for some alteration either to tip of horn or roller jewel. Consult the curve safety test before making any change.

LESSON 78

TEST LESSON NO. 22 B—CENTRAL PART OF HORN—DOUBLE ROLLER

372. *Curve Test.*—The object of this division of the curve test is to find out, if, in a double roller escapement, the central parts of the lever horn can catch or hold the roller jewel, and to learn the condition of the curve safety lock.

Method—The lever is lifted off its bank at a time to cause contact, of the central parts of the lever horn with the roller jewel. Hold the *parts in contact* and slowly rotate the balance.

The rotation of the balance causes the roller jewel to rub along the face of the horn; this rub is to be expected, but nothing resembling a catch should be felt. If a catch is detected, do not confuse a catch caused by a tripping error with a catch due to a fault in the shape of the horn, or in the position of the roller jewel. Before altering the shape of the horn in a double roller escapement make a careful inspection of the safety locks. (See curve safety test.)

LESSON 79

TEST LESSON NO. 22 C—SLOT CORNERS—DOUBLE ROLLER

373. *Curve Test.*—We employ this section of the curve test to learn if the roller jewel can catch on, or in the vicinity of the slot corners.

Method—Bring the roller jewel, and that part of the lever horn very close to the slot corner into contact. Next rotate the balance so as to bring the roller jewel into the slot. During the rotation of the balance, contact of the horn with the roller jewel will be felt; also when the roller jewel passes the slot corner a rubbing of the parts will be felt. The rubbing is permissible, but a catch is not allowable. Should the roller jewel catch on the horn, or on the slot corner make use of all tests and correct all other errors before altering the relation of the curve of the horn to the roller jewel. Both horns of course should be tested.

LESSON 80

TEST LESSON NO. 23 A—DOUBLE ROLLER

374. *Curve Safety Test.*—The purpose of the curve safety test is to determine if a safety lock of tooth on pallet is present,—

when any part of the curve of the horn *is in contact* with the roller jewel.

In a double roller escapement the moment the guard point enters the crescent, the roller jewel and horn can be brought in contact, simply by lifting the lever off its bank. When these parts are in contact the safety lock should be sound.

The safety lock in a double roller escapement should be inspected in the following positions: *F*irst, at the moment the guard point enters the crescent; secondly, from about the central part of the horn up to the slot corner; thirdly, when the roller jewel is opposite the slot corner.

Defects in the safety locks, no matter from what cause, always demand correction. Before altering the shape of any part of the horn, use other escapement tests as a confirmative that the fault is really due to a defect in the shape of the horn.

LESSON 81

TEST LESSON NO. 24 A—END OF HORN—SINGLE ROLLER

375. *Curve Test.*—The intention of this part of the curve test is to learn, if in a single roller escapement, the roller jewel can catch on the end, or central parts of the horn.

Method—Bring the guard pin in contact with the edge of the table, hold the parts in contact while the balance is being rotated so as to bring the roller jewel past the end and central part of the horn. *No* contact of the roller jewel with this part of the horn should be felt; there is, however, no real objection to the parts rubbing so long as nothing resembling a catch is detected. If a catch develops alterations should be made.

As a matter of fact, *in single roller escapements* that part of the horn from the tip to the central part could be dispensed with without jeopardizing the escapement.. The curve test in single roller escapements is of *little importance* and can usually be omitted.

LESSON 82

TEST LESSON NO. 25 A—HORN AND SLOT CORNERS—SINGLE ROLLER

376. *Curve Test.*—The object of this division of the curve test is, to determine if the roller jewel under any circumstances can catch on the slot corners, or on that part of the horn near the slot corners (single roller escapement).

Method—Rotate the balance so as to bring that part of the table's edge adjacent to the crescent, opposite the guard pin. When the parts are in the position mentioned, lift the lever off its bank, thereby causing the guard pin to touch the edge of the table. If the directions have been carefully followed, the guard pin will touch the edge of the table at a point very close to the crescent. Maintain pressure against the lever and rotate the balance just enough to bring the guard pin barely inside the crescent. The instant the guard pin enters the crescent, and because of the pressure exercised on the lever, the horn and roller jewel come into contact with each other. If the curve of the horn is adapted to the position of the roller jewel the latter will rub over the face of the horn and past the slot corner in a smooth manner, or at least without developing any undue amount of friction. No catch of the parts is permissible; if such is discovered an alteration is required. Never jump at a conclusion, but make use of *many* tests to determine the cause of a fault.

LESSON 83

TEST LESSON NO. 26 A—SINGLE ROLLER

377. *Curve Safety Test.*—The curve safety test as employed in single roller escapements is used to determine the existence of tripping errors.

Method—To learn if a tripping error is present, bring the curve of the lever horn in contact with the roller jewel after the manner described in Test Lesson No. 25 *A*; then, while the parts are held in touch with each other, examine the safety lock. Should a tripping error be located, employ other escapement tests before assigning the cause of the error to either a defect in the curve of the horn, or position of the roller jewel.

Index to Test Lessons 27 A to 29 B

EXAMINATIONS AND TEST FINDINGS

378. *Escapement Examination*—Elgin and South Bend.—

LIST 27

27 A. Escapement Examination. Error, Deep Drop Locks.
27 B. Escapement Examination. Error, Light Drop Locks.
27 C. Escapement Examination. Error, Unequal Drop Locks.

LIST 28

28 A. Proof *F*indings. Elgin and South Bend Contrasted.

LIST 29

29 *A*. South Bend Escapement. Error, Light Locks.
29 B. South Bend Escapement. Error, Deep Locks.

LESSON 84

COMPARISONS OF THE TESTS

TEST LESSON NO. 27 A—DEEP DROP LOCKS—ELGIN TYPE

379. *Examining an Escapement.*—

Remarks—The three following lessons are designed to make you familiar with the full routine of escapement testing. You will find it to your advantage to *compare* similar test findings in all *three* lessons. Of course, when a comparison is made bear in mind *the difference in the locks.* Although the routine is apparently extensive enough to consume a great deal of time, as a matter of fact, once you are acquainted with the tests, two to five minutes will decide if changes are necessary. That it is not always necessary to follow out the full routine of testing can be learned by studying "Bench Problems." One soon becomes familiar with cause and effect and can then quickly locate any defect liable to cause escapement trouble. In the previous "Test Lessons" each "test" is *separately treated*—here we have them *combined.*

It is assumed that the escapements we are about to examine *were originally perfect,* but some one *changed their respective locks* into *deep, light* and *unequal.* The length of the levers, the position of each roller jewel, the location of the guard points, and sizes of tables *are exactly right.* When the drop locks *are matched* to the respective length of the levers, *each escapement* will then belong to that class we have designated correct, or at least commercially correct. The student should closely study Lesson No. 87, in which the escapement types and tests are contrasted.

Condition of Escapement—Drop locks *deep* but equal. Corner freedoms *excessive.* Guard freedoms *excessive.*

Angular Test Findings (Elgin)—The angular test shows that each tooth has too much contact with its pallet stone as illustrated by Fig. 31.

Corner Test Findings, Banked to Drop (Elgin)—By the corner test we discover an excess of freedom between the roller jewel and each slot corner. (Corner freedom.)

Corner Safety Test Findings (Elgin)—When the slot corners are held in contact with the roller jewel there is an overabundance of safety lock.

Guard Test Findings, Banked to Drop (Elgin)—By the guard test a surplus of freedom is found between the guard point and edge of table. (Guard freedom.)

Butting Test Findings—There is no evidence of a tendency of the guard point to butt or stick against edge of table.

Guard Safety Test Findings—An *excessive* amount of safety lock is noticeable.

Curve Test Findings—This test shows that the relationship of the horns to the roller jewel is normal.

Curve Safety Test Findings—Contact of the horn with the roller jewel shows as in the other safety tests, a *great* amount of safety lock.

Draw—The condition of the draw is satisfactory.

Drop, Inside and Outside—On account of the deep locks the extent of the drop is curtailed.

Shake, Inside and Outside—As we discovered a lessened amount of drop, we correspondingly find the shakes dangerously decreased. (Students should learn to realize the importance of shake.)

Tooth and Pallets' Impulse Planes—The action of each tooth over each pallet jewel's impulse face is fairly satisfactory.

LESSON 85

COMPARISONS OF THE TESTS—TEST LESSON NO. 27 B—DROP LOCKS LIGHT—ELGIN TYPE

380. Note.—The assumption is that this escapement was correct when it left the factory, but afterwards some one made the drop locks lighter.

Condition of Escapement—Drop locks *light* but equal. Corner freedoms *deficient.* Guard freedoms *deficient.*

Angular Test Findings (Elgin)—The angular test shows each tooth as discharged from its pallet jewel. (See drawing No. 30.)

Corner Test Findings, Banked to Drop (Elgin)—The corner test shows a want of the correct amount of freedom between the slot corners and roller jewel. (Corner *F*reedoms.)

Corner Safety Test Findings—A deficiency of safety lock is observed, when the slot corners are held in contact with the roller jewel.

Guard Test Findings, Banked to Drop (Elgin)—We discovered by the guard test that there is a want of freedom between the guard point and edge of table. (Guard freedom.)

Butting Test Findings—No tendency of the guard point to butt or stick against the edge of the table is found.

Guard Safety Test Findings—The extent of the safety locks are less than desirable.

Curve Test Findings—The curve test does not show any irregularities of the horn with the roller jewel.

Curve Safety Test Findings—When the horns and roller jewel are in contact, we again find the safety locks are deficient.

Draw—The condition of the draw is satisfactory.

Drop, Inside and Outside—The amount of drop, both inside and outside, is somewhat excessive.

Shake, Inside and Outside—The shakes, inside and outside, show plenty of freedom.

Tooth and Pallet's Impulse Plane—The action of the escape wheel teeth over the impulse planes of the pallet jewels are satisfactory.

LESSON 86

COMPARISON OF THE TESTS—TEST LESSON

NO. 27 C—DROP LOCKS UNEQUAL—ELGIN TYPE

381. *Examining an Escapement.*—

NOTE—Originally this escapement was perfect, but somebody tampered with the drop locks, making them unequal; consequently the escapement is out of angle.

Condition of Escapement—Drop locks *unequal.* Corner freedom, on one corner *excessive.* Corner freedom, on opposite corner *deficient.* Guard freedom on one side *excessive.* Guard freedom on opposite side *deficient.*

Angular Test Findings (Elgin)—As the drop locks are unequal, the angular test shows dissimilar positions of contact of each tooth with each pallet; or perhaps, one tooth shows too much contact, while the other tooth is discharged from the opposite pallet jewel.

Corner Test Findings, Banked to Drop (Elgin)—The corner test shows an excessive amount of corner freedom on one corner, while on the opposite corner the amount of corner freedom is very deficient, or at least we find that the corner freedoms are unequal.

Corner Safety Test Findings—On one corner we find a surplus of safety lock. The opposite corner shows a shortage of safety lock.

Guard Test Findings, Banked to Drop (Elgin)—The guard test reveals too much freedom on one side of the table, and too little on the oposite side.

Butting Test Findings—A butting error is not discoverable.

Guard Safety Test Findings—We find on one pallet jewel an excessive amount of safety lock. On the opposite pallet the safety lock is deficient or wanting.

Curve Test Findings—No defects of curve of horn with roller jewel are found by this test.

Curve Safety Findings—When the roller jewel is brought into contact with one curve of the horn a great amount of safety lock is found. A similar test, with the opposite curve of horn, shows a shortage of the safety lock.

Draw—The draw is fairly satisfactory.

Drop, Inside and Outside—The drops are unequal—viz., the inside drop differs in extent from the outside drop.

Shake, Inside and Outside—The shakes, like the drops, are also unequal. The lesser of the two shakes probably show a dangerous lack of freedom.

Tooth and Pallets' Impulse Plane—The action of each tooth over the impulse faces of the pallet jewels will perhaps be fair.

LESSON 87

ESCAPEMENT CORRECT—TESTS AND TYPES

CONTRASTED—TEST LESSON NO. 28 A—PROOF FINDINGS

382. *Elgin and South Bend Proof Findings.*

Angular Test, *Proof Findings, Elgin* *Type*—In the *Elgin* type of escapement the angular tests proof findings show a slight *contact* of each tooth with each pallet. (See Fig. 29.)

Angular Test, *Proof Findings, South Bend* *Type*—The *South Bend* escapement will show results *contrary* to above—namely, the escape wheel teeth are *discharged* from their respective pallet jewels. (See Fig. 30.)

Corner Test, Proof Findings, Elgin *Type, Banked to Drop*—The proof findings of the corner test will show when the escapement is banked to drop, a certain amount of freedom between the roller jewel and slot corners. (Elgin type.)

Corner Test, *Proof Findings, South Bend Type, Banked to Drop*—When the escapement is banked to drop the proof findings show *contact* of the roller jewel with the slot corners. (South Bend type.)

Guard Test, *Proof Findings, Elgin* *Type, Banked to Drop*
The proof findings of an *Elgin* type of escapement when banked to drop shows freedom between the guard point and table.

Guard Test, *Proof Findings, South Bend* *Type, Banked to Drop*—Contact of the roller jewel with the slot corners when the escapement is banked to drop is the correct proof findings for *Dueber* and *South Bend* escapements.

In this "Lesson" we have contrasted the correct or proof findings of each test for escapements of the Elgin and South Bend types. Their distinctive differences must be recognized.

LESSON 88

TEST LESSON NO. 29 A—ERROR, LIGHT LOCK

383. *South Bend Escapement, Comparison of Tests.*

NOTE—The escapement in this lesson is assumed to be of the *South Bend* type. It possesses but one real error—namely, the drop locks are unsafely light. The effect of unsafe locks is shown by the various tests. A comparison of the tests in this and the following lesson will be found profitable if we bear in mind the opposite defects in the locks—viz., one is light, and the other deep. The contrasting differences should be remembered.

Angular Test (South Bend)—The teeth of the escape wheel remain in contact with their respective pallet jewels. This is contrary to the proof findings for this type of escapement.

Corner Test, Banked to Drop (South Bend)—The drop locks are light, and to meet the requirements of this test the escapement is banked to drop; combinedly these conditions prevent the roller jewel from either entering or leaving the slot, consequently the corner test cannot be made.

Guard Test, Banked to Drop (South Bend)—To make the guard test, it is also necessary to bank the escapement to drop, as the drop locks are light, the banking pins stand closer together than they should. The result is the guard point is jammed against the edge of the table, which makes the test unsatisfactory.

LESSON 89

TEST LESSON NO. 29 B—ERROR, DEEP LOCKS

384. *South Bend Escapement—Comparison of Tests.*—

NOTE—The error in this escapement is the drop locks are deep. We again assume that this is the only defect, the remaining escapement parts being perfect.

Angular Test *(South Bend)*—The error of deep locks is recognized by the fact that the escape wheel teeth show an over-contact—viz., too much contact of the teeth with their respective pallet jewels. (See Fig. 31.)

Corner Test, *Banked to Drop (South Bend)*—The error as shown by the corner test is that freedom is found between the slot corners and roller jewel, freedom being an incorrect corner test finding for this type of escapement.

Guard Test, *Banked to Drop (South Bend)*—The guard test, like the corner test, shows freedom between the guard point and table. The finding is incorrect for this type of escapement.

The foregoing lessons on South Bend escapements should be studied in connection with our previous instructions. This will place students in command of the main features controlling escapements of this type.

BENCH PROBLEMS

Index to Bench Problems Nos. 1 to 14

385. 1. Locks deep. Lever Long.
2. Locks light.
3. Butting error.
4. Lever long. Table small.
5. Defective entry of tooth onto pallet's impulse face.
6. Want of inside shake.
7. Deep locks. Guard pin radius long. Lever long.
8. Out of angle. Lever bent
9. Out of angle. Locks irregular.
10. Out of angle. Locks deep and irregular. Lever bent.
11. Locks deep and irregular. Draw bad.
12. Deep locks. Curve of horns.
13. Light locks and a corner trip.
14. Deep locks. Long lever. Table large.

LESSON 90

The following lessons under the title of bench problems have been taken from actual experiences at the bench. As the lessons contain nothing artificial, their practical worth from an instructive standpoint is increased.

386. *Bench Problem No. 1.*

Errors—Locks deep. Lever long. Elgin type. Single roller.

SECTION *A*

Remarks (1)—An inspection of the *total* locks on each pallet proved them deep, but equal.

SECTION B

Alteration (2)—Banked escapement to drop.
Remarks (3)—Found the drop lock deep on each pallet.

SECTION C

Alteration (4)—Decreased the drop lock on each pallet, and rebanked escapement to drop.

SECTION D

Remarks (5)—Learned that the roller jewel scraped past each slot corner, passing them with difficulty.

SECTION E

Alterations (6)—Filed away a part of the horns and slot corners and repolished same.

Remarks (7)—The corner test now showed correct freedom between roller jewel and each slot corner. The angular test proved the parts well matched.

SECTION F

Alteration (8)—Spread the bankings for slide.

Remarks (9)—The changes made greatly improved the escapement. In this instance the readjustment of the guard pin to the table gave no trouble.

LESSON 91

LIGHT LOCKS—ELGIN TYPE—SEVEN JEWELED

387. *Bench Problem No. 2.*

SECTION A

Remarks (1)—The watch was recently cleaned, but stopped while on the rack.

SECTION B

Alteration (2)—Banked escapement to drop.

Remarks (3)—Found the locks unsafely light for this grade of watch. No freedom was discovered by either guard or corner tests. The lack of guard and corner freedoms alone prevented tripping.

SECTION C

Alterations (4)—Slightly increased the drop locks and re-banked to drop.

Remarks (5)—Result of altering the locks. The guard and corner freedoms became correct, and the angular test showed the lever's length as matching the locks.

SECTION D

Alteration (6)—Opened the bankings for slide.

Remarks (7)—The watch gave no further trouble.

LESSON 92

BUTTING ERROR—ELGIN TYPE—SINGLE ROLLER

388. *Bench Problem No. 3.*—

SECTION A

Remarks (1)—The owner complained about the irregular timekeeping qualities of the watch.

SECTION B

Remarks (2)—The locks appeared correct. By the angular test the parts were well matched. The corner test showed correct corner freedom. The guard test showed the guard pin inclined to stick or butt against edge of table.

SECTION C

Alterations and Remarks (3)—Banked escapement to drop, then tried adjusting guard pin. Succeeded in overcoming the butting, but the guard freedom was now deficient.

SECTION D

Alteration (4)—Placed the table in lathe, and slightly reduced its size.

Remarks (6)—Replaced all parts in position and found that the guard freedom was correct. The butting error had disappeared.

SECTION E

Alteration (7)—Opened the bankings, adding slide.

Remarks (8)—Watch very satisfactory.

LESSON 93

SINGLE ROLLER—ELGIN TYPE—ERRORS, LONG LEVER AND ROLLER TABLE TOO SMALL

389. *Bench Problem No. 4.*—

SECTION A

Remarks (1)—When the watch was received for repairs its apparent condition was fair, but the balance had a poor motion.

Alteration (2)—Banked escapement to drop.

Result (3)—Found the drop locks correct.

SECTION B

Remarks (4)—Replaced balance, placing the roller jewel in the slot.

Result (5)—Learned that the roller jewel was unable to leave the slot.

Remarks (6)—Two facts are plain—(a) the drop locks are correct; (b) the roller jewel is held in the slot.

Inference (7)—The lever is too long.

SECTION C

Alteration (8)—Reduced the lever's acting length enough to allow the roller jewel to pass out of the slot.

Tests and Alteration (9)—Made use of the corner test to still further reduce the lever's acting length and secure the correct amount of corner freedom. The length of the lever was confirmed by the angular test.

SECTION D

Remarks (10)—With the guard pin straight too much freedom existed between guard pin and table.

Alteration (11)—The guard pin was made question mark in shape to secure the correct guard freedom.

SECTION E

Alteration (12)—The addition of slide then placed the escapement in good running order.

LESSON 94

ERROR, DEFECTIVE ENTRY OF TOOTH ONTO PALLET'S IMPULSE FACE

390. *Bench Problem No. 5.*—

Remarks (1)—The owner of this watch complained that it gave very poor service. Every six months or so it was in some repair shop. After repairing—which the owner said "usually was termed cleaning"—it rendered fair service, but soon deteriorated.

The Tests (2)—The tests showed a normal condition of the parts.

The Defect (3)—The defect present was due to the irregular action of the lifts of tooth and pallet. In this escapement, when *unlocking* took place, the *pallet corner* acted on the impulse face of the tooth.

Remarks (4)—When the lifting actions are correct (see former explanations) the *tooth corner* slips on to the pallet jewel's impulse face.

Alteration (5)—Changed the lifts to conform with Remarks (4).

Remarks (6)—After a year's service the owner reported the watch as "excellent."

LESSON 95

ERROR, WANT OF INSIDE SHAKE

391. *Bench Problem No. 6.*—

Remarks (1)—Some watch repairer had lately overhauled the watch. Since that time, as the owner expressed it, the watch had "stopping fits."

Inspection (2)—The balance was removed and, while inspecting the locks, it was noticed that the drops were irregular. The inside drop and inside shake being deficient, especially the latter.

Alteration (3)—As the pallet jewels bore evidence that they had been tampered with, a change was made. The locks were lessened and pallets spread slightly.

Remarks (4)—The effect of alterations were: A correct lock, and the drops equalized, besides a safe amount of shake inside and outside was secured. Result stoppage did not again take place.

LESSON 96

SINGLE ROLLER — ELGIN TYPE — ERRORS, DEEP LOCKS, RADIUS OF GUARD PIN LONG, LEVER LONG

392. *Bench Problem No. 7.*—This was an old watch of cheap construction and low grade. It was given to a beginner in escapement work, the instructions being to put same in order. The following took place:

The Locks—The locks according to the student's observation were exceptionally deep.

The Angular Test—The student next used the angular test, *neglecting* the precaution of bending the guard pin *away* from the table. This test showed each tooth as discharged from each pallet. The student then hastily assumed the discharge of the teeth implied that the lever's acting length was long. Acting on this supposition, the lever's length was made shorter, and thereby totally ruined.

Escapement's Real Condition—The *total* locks, as afterwards determined, were excessively deep. The drop locks were correct in amount, the *excess* being entirely attributable to slide.

According to previous instructions, when the angular test is used it is wisest for beginners to bend the guard point out of the way. This student failed to follow instructions and, as the guard point was too far forward, it came into contact with the table, helped move the lever, and in consequence the angular test showed the teeth as discharged.

For want of a little precaution a new lever had to be fitted. *You* will avoid trouble and erroneous conclusions by banking every escapement to drop that you desire to test, and until thoroughly familiar with the angular test *remove* the guard pin from the table.

LESSON 97

OUT OF ANGLE—CAUSE, BENT LEVER

393. *Bench Problem No. 8.*—

SECTION *A*

Alteration (1)—Banked escapement to drop.

Remarks (2)—Result of Alteration (1). The locks were found to be irregular, one lock being greater than its fellow.

Inspection (3)—Inspected the lever and discovered it was bent.

SECTION B

Alterations (4)—Without further examination the lever was made straight, and escapement was rebanked to drop.

Remarks (5)—Straightening the lever equalized the drop locks and placed the escapement "in angle"

SECTION C

Alteration (6)—Slide, a very necessary feature of every normal escapement, was then added.

LESSON 98

OUT OF ANGLE—LOCKS IRREGULAR—ELGIN TYPE

394. *Bench* Problem *No. 9.*—

SECTION A

Alteration (1)—Banked escapement to drop.

Remarks (2)—*F*ound drop locks irregular. The lock on one stone exceeding the lock on opposite stone.

SECTION B

Remarks (3)—Replaced balance.

Remarks (4)—Rotated balance and learned that the roller jewel could pass one slot corner, but was unable to get past the opposite corner.

Remarks (5)—We have discovered two things—drop locks irregular and a defective relationship of the roller jewel with the slot corner. The indications point to the escapement as out of angle.

SECTION C

Test (6)—Tried the angular test. To do so had to open out one banking—see Remarks (4). Result, contact of one tooth with its pallet jewel. The opposite tooth was discharged. Here we again find additional evidence that the escapement is out of angle.

Remark (7)—All tests unite in declaring the escapement is out of angle.

SECTION D

Alteration (8)—Made the drop locks equal, and rebanked escapement to drop.

Remarks (9)—Equalizing the locks placed the escapement in angle, as shown by corner, guard, and angular test.

SECTION E

Alteration (10)—Spread the bankings to provide slide.

LESSON 99

OUT OF ANGLE—LOCKS DEEP AND IRREGULAR—LEVER BENT—ELGIN TYPE

395. *Bench Problem No. 10.*—

SECTION A

Alteration (1)—Banked escapement to drop.

Remarks (2)—Personal inspection of the locks showed they were deep and irregular. Also the lever is bent.

SECTION B

Alterations (3)—Straightened the lever and rebanked escapement to drop.

Remarks (4)—The locks still remain deep and irregular.

SECTION C

Test (5)—Tried the guard test and found the guard freedoms unequal.

Test (6)—By the corner test, the corner freedoms were proven to be unequal.

Remarks (7)—The irregular locks and the unequal corner and guard freedoms all confirmed the escapement as being out of angle.

SECTION D

Alteration (8)—The pallet stones were reset, making drop locks lighter.

Alteration (9)—Resetting the pallet stones necessitated banking the escapement to drop.

Remarks (10)—The drop locks are now correct. The corner, guard, and angular test all prove that the escapement is satisfactory.

SECTION E

Alteration (11)—The banking pins were next adjusted to provide the necessary slide.

LESSON 100

ERRORS—LOCKS AND DROPS IRREGULAR —DRAW BAD

396. *Bench Problem No. 11.*—"For some months the watch had been out of a repair shop," so the owner said. It was brought in for want of accuracy, its timekeeping qualities being poor.

When the escapement was banked to drop, a difference in the locks and the drops was noticed. On the entering pallet the draw was very poor, as proven by the fact that the lever when lifted away from its bank would remain in position, showing no tendency to return to its bank. Whenever in the course of usage, a shock threw the lever off its bank contact of the guard point and table was inevitable. This, of course, spoiled the timekeeping.

To correct this defect the entering pallet stone was given a greater slant, its end being tilted away from the opposite pallet. This change greatly improved the draw, the locks, and the drops, besides improving the action of the tooth over the pallet's impulse face.

LESSON 101

SINGLE ROLLER—ELGIN TYPE—ERRORS, DEEP LOCKS—CURVE OF HORNS

397. *Bench Problem No. 12.*—This watch had seen service for probably thirty years. Its general appearance was dilapidated.

Our first observation happened to be of the extent of the lever's motion from bank to bank, the bankings being widely apart. By placing a finger on the balance and guiding it back and forth it was learned that the drop locks were deep. The dial was then removed and the excessive depth of the locks confirmed. Without further investigation the watch was taken apart and the drop locks made correct.

After correcting the locks and reassembling the watch the escapement was banked to drop and the angular test applied. As its findings were satisfactory, we considered the parts as matched.

The guard pin was found to be somewhat close to the table, but by manipulating it, correct freedom between pin and table was established.

The curve test was next tried. It was then discovered that owing to the shape of the upper parts of the horns the roller jewel would catch on their tips. To overcome this defect the end of each horn was changed in shape.

The necessity for changing the shape of each horn is attributable to the altered position of the guard pin. Slide was then provided, which completed the operation.

LESSON 102

DOUBLE ROLLER—ELGIN TYPE—ERRORS, LIGHT LOCKS AND A CORNER TRIP

398. *Bench Problem No. 13.*—

SECTION A

Remarks (1)—This watch was recently purchased. The owner returned it, complaining that its timekeeping qualities were uncertain.

Alteration (2)—Banked escapement to drop.

SECTION B

Remarks (3)—Banking the escapement to drop brought out two defects—(a) that the drop locks were unsafely light; (b) on replacing the balance no freedom whatever existed between the guard point and safety roller.

Remarks (4)—The fact that the guard point was in close contact with the edge of the table acted as a preventive of tripping.

SECTION C

Test (5)—The corner test was next employed. Result of this test: *A* slight amount of corner freedom was found.

Remarks (6)—As slight as the corner freedom was, it allowed the escapement to trip. The tripping error was evidently due to the unsafely light locks.

SECTION D

Alterations (7)—Increased the drop locks and rebanked escapement to drop

SECTION E

Test (8)—The corner and guard tests were again used, each now showed freedom and an absence of any tendency towards a tripping error.

SECTION F

Alteration (9)—The addition of slide placed the escapement in good order.

LESSON 103

SINGLE ROLLER—MOVEMENT MARKED AD JUSTED—ELGIN TYPE—ERRORS, DEEP LOCK, LONG LEVER, TABLE DIAMETER LARGE

399. *Bench Problem No. 14.*—

SECTION A

Remarks (1)—This was a new movement and was examined because of the large total lock it possessed. The total lock approached 6 degrees, as judged by the table in Paragraph No. 183.

Alteration (2)—Banked escapement to drop.

Remarks (3)—On banking to drop it was learned that the drop lock approximated 4 degrees. Evidently the slide amounted to about 2 degrees.

SECTION D

Remarks (4)—Decided to try the angular test.

Test (5)—The angular test proved the parts matched—namely, the lever's length was suited to the deep lock.

Remarks (6)—Although the angular test showed the parts as matched, the deep lock could not be allowed to go uncorrected.

SECTION C

Alteration (7)—Altered the drop lock, making same correct.

Alteration (8)—Rebanked the escapement to drop.

SECTION D

Alteration (9)—Bent the guard pin *away* from the edge of the table.

Remarks (10)—It was now discovered that owing to lessening of the locks and rebanking to drop, that only with *difficulty* could the roller jewel be brought past the slot corners.

Remarks (11)—The lack of corner freedoms as above mentioned indicated a change as necessary.

SECTION E

Alterations (12)—To provide corner freedom, cut the horn and slot corners away. Frequent use was made of both the corner and angular test to prevent an over-reduction of the lever's acting length.

Remarks (13)—In this way the correct corner freedoms were secured. The angular test also showed the new drop locks were adapted to the lever's altered length.

SECTION F

Alteration (14)—Made the guard pin straight—see Alteration (9).

Remarks (15)—The result of straightening the guard pin caused contact of pin with table.

SECTION G

Alteration (16)—As a matter of experiment, the bankings were opened.

Remarks (17)—The guard pin, it was then discovered, would stick or butt against edge of table.

SECTION H

Alteration (18)—Rebanked escapement to drop.

Alteration (19)—Placed the roller table in the lathe and ground off its edge, thereby reducing its size.

Alteration (20)—Slightly advanced the guard pin's position.

Remarks (21)—Replaced all parts in position. Found as result of Alterations (19) and (20) that the butting error had disappeared, and a correct amount of guard freedom secured.

SECTION I

Alteration (22)—The bankings were next opened for slide.

Remarks (23)—The watch was timed in positions. It afterwards proved an excellent timepiece.

Index to Hints and Helps

LESSON 104

HINTS AND HELPS NO. 1

401. *Testing and Altering a Light Lock.*—Among the cheaper grades of watches we sometimes encounter a lock that looks suspiciously light. When such is found, and an alteration is desirable it is best to make an exhaustive test covering the entire escapement action. Usually an inspection of the drop locks reveals that one lock slightly exceeds the other. Should this be the case, and an increase of the lesser lock be desirable. Try the guard safety test, or the corner safety test, or both. Before making any alteration examine and compare the extent of the respective safety locks. Do this as a guide towards deciding which is the lesser lock. As a further precaution, test the lock of many teeth of the escape wheel, for the reason that frequently in cheaper watches the length of the escape wheel teeth vary.

In low grade watches we can best determine whether a light lock is satisfactory, or unsatisfactory, by banking the escapement to drop, using the angular, corner, guard and safety tests to arrive at a definite conclusion.

LESSON 105

HINTS AND HELPS NO. 2

402. *Estimating the Three Safety Locks.*—The extent of the safety or remaining lock should be investigated and estimated in three places, as follows: First, by bringing the guard point in contact with the edge of the table (guard safety test); then, with an eyeglass, note the extent to which the tooth remains locked on the pallet jewel.

The second place for making an observation of the safety lock is, when the corner of the lever slot is brought in contact with the face of the roller jewel (corner safety test).

The third place in which the safety lock should be estimated is when the curve of the horn is brought into contact with the roller jewel (curve safety test). The contact mentioned can only be obtained when the guard point is *within* the crescent.

LESSON 106

HINTS AND HELPS NO. 3

403. *Resetting Pallet Jewel—Both Pallet Stones Out of Place—Elgin Type.*—

NOTE—The problem stated below involves resetting the pallet jewels in such a way that the drop lock will be safe, light, and adapted to the fork-roller jewel action. When this has been attained the escapement will be well matched.

Should you at any time encounter an escapement in the condition mentioned below, do not commence alterations by attempting to set the pallet jewels so the lock looks about right, because as a rule you will find the suggestions in Methods A and B much more desirable.

In connection with this lesson students are advised to apply the instructions in a practical way—namely, obtain a watch, then remove the pallet jewels out of their settings and entirely alter the position of the banking pins. This done, Methods A and B can be used and very practical lessons be learned about resetting pallet stones, and in relocating lost positions of the banking pins.

The following represents the state of the escapement:

Condition—Both pallet jewels removed from their settings. Bankings tampered with, so they are useless as guides.

Method A (Guard Test)—When both bankings have been disturbed, and both pallet stones removed, we can by employing the following system again locate the positions the bankings will occupy when the escapement is banked to drop. This position of the bankings being found, it facilitates the resetting of the pallet stones.

Directions—Place the lever and balance in position, next turn in each banking so as to bring the guard point *nearly* in contact with the edge of the table (Elgin type). Assuming that the location of the banking pins as above determined represents the position for correct drop lock, the pallet jewels are reset accordingly. If the table's diameter is correct, and the guard pin straight no trouble is likely to be experienced.

Should the table's diameter be too large the result of resetting will be a *deep* lock. On the other hand, if the table's diameter is too small the drop lock will be correspondingly *light.*

Whenever the drop locks as determined by Method A prove unsatisfactory—that is, light or deep—it is then advisable to employ Method B as explained in the following lesson.

NOTE—If the escapement is of the South Bend type, each banking pin must be adjusted so as to bring the guard pin in direct *contact* with the edge of the table. We thereby establish a correct position of the bankings, which enables us to reset the pallet stones for drop lock only.

LESSON 107

HINTS AND HELPS NO. 4

404. *Resetting Pallet Jewels—Both Pallet Stones Out of Place—Elgin Type.—*

NOTE—The state of the escapement is exactly the same as in the previous lesson.

Conditions—Both pallet jewels removed from their settings. Position of banking pins so altered they are useless as guides.

Method B (Corner Test)—As before, the problem is to reset the pallet stones so the drop lock will be light yet safe, and also adapted to the action of the roller jewel with the fork.

To make use of Method B, place the lever and balance in position, next revolve the balance so as to bring the roller jewel opposite the corner of the lever slot, then hold balance in this position and adjust each banking pin so as to establish a slight freedom between the slot corner and face of roller jewel. With this position of the banking pins as a guide the pallet stones are reset for drop lock only.

If owing to some disturbing element in either the acting length of the lever or the location of the roller jewel we obtain in place of a correct lock, a lock that is *deep*, we realize that the lever's length is too long or that the roller jewel is too far forward. Should the drop lock obtained by Method B result in establishing a drop lock that is too *light* we then know that either the lever is short, or the roller jewel is not sufficiently forward.

A failure to obtain a correct drop lock by Methods A and B mean that the escapement is mismatched and that alterations will have to be made to improve the escapement. (See following lesson.)

LESSON 108

HINTS AND HELPS NO. 5

405. *Remarks on Methods A and B.*—When Methods A and B fail us we must then depend on sight and judgment to reset the pallet jewels so as to provide a drop lock suitable for that particular escapement. With *some* assumed standard of correct lock present we then fall back upon the corner, guard and angular tests. Alterations must be made in accordance with the instructions given in the test lessons. You may feel certain that the parts are very much mismatched when Methods A and B fail to yield satisfactory results.

LESSON 109

HINTS AND HELPS NO. 6

406. *The Resetting of One Pallet Stone—Elgin Type.*—Given an escapement with one pallet stone remaining in its original position, the opposite pallet jewel being out of place, and we desire to reset the loose stone in conformity with the stone still in place, we can for this purpose make use of one or both of the following methods. No trouble will be experienced unless the escapement is mismatched; if so, the test lessons should be consulted and suitable alterations made.

Guard Test Method—Turn in the bankings so the guard point nearly but not quite touches the edge of the table. With this new position of the bankings as a guide, set the loose pallet jewel for drop lock.

Corner Test Method—Close in the bankings to such an extent that when the corner test is applied a slight freedom is found between the slot corner and the roller jewel. With the bankings established in this position the pallet stone should be reset for drop lock.

LESSON 110

HINTS AND HELPS NO. 7

407. *Correction of Guard Point and Table Errors.*—

NOTE—The purpose of this lesson is to call attention to some practical and useful point. Although some of them have been previously mentioned, the gathering together of these items will add to their usefulness.

Contact of Guard Point and Table.—When a watch of the Elgin type is banked to drop and we discover contact of the guard point with both sides of the table the remedies to be applied are as follows:

A. Replace old table with a new one of lesser diameter.

B. Continue the use of the old table, but lessen the diameter of the guard point. (Methods A and E are usually preferable.)

The suggested reduction of the thickness of the guard point is best done in single-roller escapements by means of a tool made to slip over the guard pin. The work of the tool is to thin the guard pin by shaving its front and sides. Such a tool can be made from a medium sized needle. A hole should be drilled in it slightly larger and deeper than the guard pin. After drilling, the hollow circular end of the tool should be sharpened. In order to prevent the tool clogging when in use, a part of the back as far in as the beginning of the hole should be filed away for clearance.

C. If we think it desirable to continue the use of the old table, the guard pin can be bent away from the edge of the table. The belly caused by bending can frequently be removed by the tool above described. If the belly is too great, better lessen the diameter of the old table and keep the guard pin upright.

D. The insertion of a new but slightly tapering pin at times helps to lessen the trouble.

E. The diameter of the table can be lessened by securing it in the lathe and turning it down a trifle. This will provide the required freedom. A lap for grinding the table's edge is superior to the graver.

F. If the Methods B, C, or D, *result in a butting action* of the guard point with edge of table, the butting error *must* be eliminated by the use of the remedies suggested in A or E.

LESSON 111

HINTS AND HELPS NO. 8

408. *Irregular Contact of Guard Point with Table.*—If we suspect or know that the guard point touches the edge of the table in some places and is free in others, it indicates:

A. The table is out of truth in the round.
B. The balance pivots may be bent.
C. Pivots working in holes too large for them.
D. Dirt or shellac about the edge of the table.
E. Roughness on the edge of the table. It is advisable to polish the edge of a table that appears rough.

LESSON 112

HINTS AND HELPS NO. 9

409. *Test for Table Wanting in Truth.*—First, close in one banking so the guard point *barely* touches the edge of the table; second, slowly rotate the balance; third, find out, by frequently trying the shake, where contact takes place and where the guard point has more or less freedom from edge of table; fourth, as regards the correction required, find the cause of the trouble and correct it. (Consult previous lesson.)

LESSON 113

HINTS AND HELPS NO. 10

410. *Replacing a Lost Roller Table.*—As the principles involved in the selection of new tables for a double roller escapement are identical with the principles governing the selection of a new table for a single roller escapement, we shall treat the subject from the standpoint of the latter.

Assuming we have in hand a single-roller escapement of the Elgin type whose drop locks are correct, the following is the procedure in supplying a new table. There are two points calling for attention—namely, the *diameter* or distance across the table and the *position* of the roller jewel. These are the governing features:

Operation A—Bank the escapement to drop.

Operation B—Select a table of such a size that a little freedom or space is left between the guard pin and edge of table. (The guard freedom of the guard test.)

Operation C—The position of the roller jewel must be such that a little freedom is found between roller jewel and the slot corners. (Corner freedom.)

To confirm the correctness of the table's dimensions and position of roller jewel, make use of all the safety tests and also employ the angular test.

Should there exist any tendency of the guard point to butt or stick against the edge of the table, the test lesson on a butting error should should be consulted. We wish also to call your attention to the fact that we stated in our opening paragraph that the drop locks were correct, hence if you discover any defects in the locks make the necessary alterations previous to fitting a new table.

LESSON 114

HINTS AND HELPS NO. 11

411. *Setting a Roller Jewel to Correctly Match the Escapement.*—When replacing a lost or loose roller jewel, pay careful attention to *size* and *position.* A roller jewel should be fitted to the fork slot and *not* to the hole in the table. Select a jewel allowing of a slight side play in the slot. If the side play is too great the lever will "jump." This jumping of the lever is easily detected by placing a finger on the balance rim, and slowly guiding the roller jewel into the slot. A roller jewel which allows the lever to jump is either too small for its slot or is loose. If too small it must be replaced by a wider jewel, because a roller jewel too small for its slot deranges the entire escapement action.

The position the roller jewel occupies in the table is of great importance. The holes in many tables are too large. When we encounter tables with large holes we must, in order to obtain the correct location of the roller jewel, be guided by the findings of the angular and corner tests, the latter test being used with the escapement banked to drop.

If we desire to set a roller jewel in an Elgin escapement of a "correct type," it is necessary to provide a little freedom between the slot corners and the roller jewel, as required by the corner test under banked to drop conditions. In this way a correct roller jewel radius is obtained.

If we bank an escapement of the South Bend type to drop, then, according to the rules controlling this type of escapement, when a roller jewel is correctly placed the corner test will show that the roller jewel just *touches* the slot corners as it passes in or out of the slot. This distinction between the escapement types must be borne in mind.

LESSON 115

HINTS AND HELPS NO. 12

l12. *The Corner Test in Practice.*—The following method of
; the corner test possesses many practical advantages. The
'iption given applies to the test in two forms—first, banked
'op; second, *not* banked to drop. When it is undesirable to
rb the position of the bankings the latter method proves very
ıl.

Րhe amount of corner freedom when an escapement is *not*
ed to drop always exceeds the amount of corner freedom
ı the escapement *is* banked to drop. The excess is of course
butable to the presence of slide.

Vhen using this test *without* banking the escapement to
if a seeming surplus of corner freedom is experienced, the
ıt of slide should be investigated, and mental deductions
of the slide from the excessive corner freedom.

'f we discover a *great deal* of corner freedom when the slide
t excessive, we realize that something is wrong and correc-
are called for. Anything of this nature that is found must
.vestigated, and corrections if they are necessary should be
. As an added help in the detection of errors relating to
subject it will be found useful to compare the amount of
l pin freedom with the corner freedom. These should be
; equal, as elsewhere explained.

Jorner Test Not Banked to Drop—First—Revolve the balance
to bring the roller jewel well past the end of the horn, and
it there.

Second—Place a watch oiler or other fine tool against the
of the lever in such a way that the lever is securely held
st its bank.

'hird—Remove finger from balance.

'ourth—Result of releasing the balance is, the roller jewel
diately bounds into slot.

'ifth—When the roller jewel settles against the opposing
of the slot, again place your finger on the balance rim, being
ıl not to move the balance the slightest.

Sixth—While the balance is being held steady and secure,
;e the watch oiler, or other fine tool, to the opposite side of

the lever, and try how much you can lift the lever away from its bank.

Seventh—The extent the lever can be moved away from its bank represents the amount of freedom present between the slot corner and the roller jewel.

Corner Test Banked to Drop—The above instructions should be followed out in detail. The only difference experienced will be that with the escapement banked to drop a *lesser* amount of corner freedom is found.

LESSON 116

HINTS AND HELPS NO. 13

413. *Calculating Dimensions of a New Escape Wheel.*—

English Ratchet Tooth Escape Wheel—If an escape wheel with ratchet shaped teeth is lost, the size of a new one is calculated as follows:

Rule—(a) Measure the distance of centers between hole for pivot of escape pinion and hole for pivot of pallet staff.

(b) Multiply the distance the center of these holes are apart by .866. The product is the *radius* of the new wheel.

Example—If the distance of centers equals 3.6 millimeters determine the *radius* of an escape wheel with ratchet teeth adapted to this distance. Following the rule stated above·

$$3.6 \times .866 = 3.11 \text{ millimeters.}$$

As answer shows the *radius* of the new wheel will measure 3.11 millimeters.

If we take a depthing tool and adjust the points to 3.11 millimeters apart and place one point of the tool in center of wheel, the other point will touch the tips of the teeth, provided the wheel is correct in size.

Fine measuring tools are desirable, such as guage 1/100 of a millimeter or 1/1000 of an inch. Most watchmakers own tools registering 1/10 millimeters, and on the principle that some idea of size is preferable to none we suggest they use such guages for measuring center distances and calculating sizes of escape wheels. A few practical experiments will demonstrate that calculations are not at all difficult.

Club Tooth Escape Wheels—The selection of an escape wheel with club teeth requires two sets of calculations, both easily made.

If the reader turns to Fig. 10, he will see that to draft a club tooth escape wheel requires an inner and an outer circle. Upon the inner circle the locking corners of all teeth will rest. Upon the outer circle the points of the teeth come in contact. To determine the size of a club tooth escape wheel we must figure the *radius* of each circle as it relates to their common distance of centers.

Radius of Inner Circle—Rule A 1—Multiply the measure of

the distance of centers by .866. The product will be the radius of the inner circle.

Radius of Outer Circle—Rule—Multiply the measure of the distance of centers by the modulus representing the angle of lift on tooth. (See following table.)

Moduli for Radius Outer Circle

Lift.	Modulus.
3°	— .892
3¼°	= .894
3½°	= .896
3¾°	— 898
4°	— .900

Question—Given a distance of centers, as 3.6 millimeters, and lifting angle on tooth as 3 degrees, (a) determine the *radius* of the inner circle (C. C., Fig. 10); (b) also determine the *radius* of the outer circle. (00, Fig. 10.)

Calculating Radius Inner Circle—

Center distance 3.6
Modulus .866
3.6 × .866 = 3.11

Radius of inner circle measures 3.11 millimeters.

Calculating Radius of Outer Circle—

Distance of centers 3.6
3° of lift. Modulus .892
3.6 × .892 = 3.21

The *radius* of the outer circle equals 3.21 millimeters.

In this manner the dimensions of any club-tooth escape wheel may be worked out. The modulus for any lift can be learned by turning to a table of tangents—for instance, the tangent of 3 degrees is .05241, divide this by 2 and we get .02620, adding it to .866 gives .892, as the modulus, which when multiplied by the distance of centers tell us the *radius* of the outer circle.

If the lift is 4 degrees its tangent is .06993, dividing by 2 yield .03496, adding this to .866 produces .900, the last modulus in the foregoing table.

Checking Sizes of Escape Wheels in Drawings or Models—The method just explained can be used to practical advantage in verifying the size of an escape wheel in a drawing or in an escapement model.

LESSON 117

HINTS AND HELPS NO. 14

414. *Replacing Lost Pallets.*—Owing to the extent of this topic, we shall confine our remarks to a few brief rules capable of easy application.

Replacing Equidistant Pallets—To obtain the measure of the distance between pallet center and locking corner of each pallet jewel:

Rule—Multiply the measure of the distance between pallet and balance centers by .5. The answer will represent the measure of the distance between pallet center and locking corner of each pallet jewel.

Example—Calculate for equidistant pallets the distance separating the pallet center to the respective locking corners of each

$$5. \times .5 = 2.5 \text{ millimeters}$$

The *locking corner* of each pallet jewel will be 2.5 millimeters from the pallet center.

To find the measure of the distance separating the letting-off corners of each pallet jewel from center of pallet staff would involve us in questions of width and lift, subjects too great for present consideration.

Replacing Circular Pallets—Without full knowledge of widths and lifts it is not possible to calculate the measure of the distance between pallet center and either the locking or letting off corners of circular pallets. What is possible and easy to follow is to calculate the measure of the distance from the pallet center to a point *midway* between locking and letting off corner of either pallet stone—viz., to center of pallet jewel. We can make use of such information as a guide. This and the practical knowledge outlined in the lessons will solve the problem of supplying a new lever.

Rule—Measure the distance of centers between pallet and balance staff and multiply same by .5.

Example—If in an escapement with circular pallets the distance of centers is 5. millimeters, what is the measure of the distance between the pallet center and center of each pallet jewel?

$5. \times .5 = 2.5$ millimeters.

The center of each pallet stone will be 2.5 millimeters from the pallet center. Students possessing drafts of the escapement can apply practically the rules given.

Index to Facts Practical and Theoretical

LESSON 118

FACTS PRACTICAL AND THEORETICAL NO. 1

416. *Angular Motion of Lever and Impulse Angle of Roller Jewel.*—The motion of the lever from bank to bank is known as the lever's angular motion. It is composed of the locks and lifts. The extent of angular motion is least when an escapement is banked to drop and greatest when slide lock is present. It is more desirable to make calculations when the extent of angular motion is least. The angular motion of the lever is a varying quantity, usually from 10 degrees to 12 degrees measured from the pallet center.

Roller Jewel's Impulse Angle—When the roller jewel during its routine of rotation meets the slot it remains for a short period in contact with the same. The extent of contact of the roller jewel with the slot is known as the roller jewel's angle of impulse. As the center of the balance corresponds with the center of the circle described by the roller jewel, the degrees of contact of roller jewel with fork slot are measured from the balance center. Generally speaking, the impulse angle of the roller jewel varies from 28 to 48 degrees. In double roller escapements from 28 to 35 degrees represents the impulse angle. The lesser the impulse angle the greater the detachment of the roller jewel from the fork. It can therefore be understood why a *well constructed* double roller escapement possesses an advantage over single roller escapements.

LESSON 119

FACTS PRACTICAL AND THEORETICAL NO. 2

417. *A Practical Method of Estimating Degrees of Lever's Angular Motion, the Locks, the Lifts and Impulse Angle of Roller Jewel.*—In a simple manner—namely, by means of a protractor—we can with a fair amount of accuracy determine the number of degrees contained in the lever's angular motion, the impulse angle, and separately the locks and lifts. Experiments of this nature are recommended because they impress the student's mind in a very practical way with facts regarding the origin and relationship of the various angles.

The Protractor—For the purpose of these experiments, attach a short upright pin exactly at the *center* of the protractor. This pin should extend about one-fourth of an inch above the surface of the instrument, the tip of the upright to be so formed that its point fits snugly into the cup or oilsink of the pallet jewel. If we shorten the pivot of the pallet staff the point of the upright pin can be shaped into a stubby cone pivot capable of entering the pallet jewel hole. By this latter method greater accuracy in the measurement of the degrees is possible.

The Escape Wheel—To prepare the escape wheel so the degrees of lift on the pallet's impulse face can be measured necessitates filing the lift from one tooth. The cutting should extend as far as the tooth's locking corner. When finished the tooth will be wedge shaped, resembling a tooth of an escape wheel having ratchet teeth.

The Lever—To the lever bar we must attach an index pointer of brass wire, shaping it so it clears the bed plate, and long enough to reach the degree marks on the protractor. If the escapement is to be kept for the purpose of demonstrating the lifts and locks, the index arm can be soldered to the lever bar. This makes the arm rigid. Should we not desire to sacrifice the lever it will be necessary to find another way of attaching the arm to the lever bar, utilizing the guard pin to steady the pointer.

Testing Lever's Angular Motion—For experiment, a 16-size bridge model is desirable, because the parts are visible and accessible. The index arm being attached to the bar, place the lever and escape wheel in the movement. As the angles relating to the motion of the lever are measured from the pallet center we place the cup or oilsink of the pallet jewel on top of the upright point attached to the center of the protractor. Place the lever against its bank and note the degree mark covered by the index arm; next shift the lever to its opposite bank and count the number of degrees the index pointer passed over (banked to bank). The number of degrees thus counted represents the lever's angular motion.

Measuring the Lock—To measure the degrees of lock, place the lever against its bank, then note the degree mark the index arm stands over; next move the lever so the tooth of the escape wheel is brought down to the lowest locking corner of the pallet jewel; again read the degree scale and thereby determine the degrees of lock. In this manner we can estimate both the slide and drop lock.

Measuring the Total Lift—The degrees of lift on tooth and pallet combined can be found by passing their combined lifting planes over each other.

Measuring Lift on Pallet—The number of degrees of lift on the impulse face of the pallet can be ascertained by passing the ratchet tooth which we prepared for this purpose over the pallet jewel's lifting plane.

Measuring Lift on Tooth—The lift on the pallet subtracted from the *total* lift of pallet and tooth will give the degrees of lift on the tooth.

Measuring the Impulse Angle—To estimate the number of degrees of the impulse angle, prepare a point for the protractor which closely fits into the recess in the cap jewel of the balance. It is also necessary to insert into a screw hole of the balance rim an index pointer of sufficient length to reach the degree scale of the protractor. It is then an easy matter to read from the scale the degrees of contact of the roller jewel with the fork slot either with the escapement banked or not banked to drop. In this way a working knowledge of the angles we have been considering can be obtained.

LESSON 120

FACTS PRACTICAL AND THEORETICAL NO. 3

418. *Proportional Method of Calculating the Lever's Acting Length and Roller Jewel Radius.*—The angular motion of the lever, and the impulse angle of the roller jewel bear a very close relationship to each other as regards their distance of centers, the acting length of the lever, and the roller jewel radius.

If we are given the lever's angular motion in degrees, and also the degrees representing the impulse angle of the roller jewel we can closely approximate the length of the lever and the theoretical radius of the roller jewel. If the lever's angular motion is 10 degrees and the impulse angle of the roller jewel is 30 degrees, the ratio of the angle is 10 to 30 or 1 to 3. The ratio of 1 to 3 approximately indicates that for every three millimeters or parts contained in lever's acting length, the radius of the roller jewel should contain one. If the lever's acting length is 6 millimeters, the roller jewel radius, according to statement just made, should be 2 millimeters.

If we are given both angles and the lever's acting length we can by proportion approximate the radius of the roller jewel.

Example—The angles are 12 and 48 respectively, the lever's acting length is 8 millimeters; calculate therefrom the roller jewel's theoretical radius—by proportion 48 : 12 : 8, multiplying the second and third terms together and dividing by the first we obtain 2 millimeters as the theoretical radius of the roller jewel.

Given the radius of the roller jewel as 1.8 millimeters, and the angles as 10 and 35, calculate the lever's acting length. As before, we make use of proportion:

$$10 : 35 :: 1.8$$
$$35. \times 1.8 = 63.0$$
$$63.0 \div 10 = 6.30$$

The lever's acting length is 6.3 millimeters.

If we know the lever's acting length and the theoretical radius of the roller jewel we can closely approximate the ratio of the lever's angular motion to the roller jewel's impulse angle.

The rule is, divide the radius of the roller jewel into the length of the lever.

Example—The acting length of a lever is 4. millimeters and the theoretical radius of the roller jewel is 1.33 millimeters—determine the ratio of their angles.

$$4 \div 1.33 = 3$$
$$1.33 \div 1.33 = 1$$

The ratio of the angle is approximately 1 to 3 or as 10 to 30.

LESSON 121

FACTS PRACTICAL AND THEORETICAL NO. 4

419. *Theoretical and Practical Radius of the Roller Jewel.*—The practical radius of the roller jewel always exceeds the roller jewel's theoretical radius. Increasing the theoretical radius by from 5 to 8 per cent. should, if the escapement is well planned, give the length of the practical radius.

If we construct an escapement and allow the roller jewel only its theoretical impulse radius, the roller jewel will, when in action, strike exactly on the slot corner. To overcome this defect the radius is made longer, thereby enabling the roller jewel to strike the wall of the slot and to conserve its action entirely within the slot.

LESSON 122

FACTS PRACTICAL AND THEORETICAL NO. 5

420. *Given Degrees of Lever's Angular Motion, and of Impulse Angle to Calculate the Theoretical and Practical Radius of Roller Jewel and the Lever's Acting Length.*—In order to follow out the below calculations, a book on trigonometry, containing tables of signs, is necessary—this, and a knowledge of multiplication and division of decimals, is all that is required to solve like problems. The system here outlined will be found more accurate than the preceding instruction by the proportional method, but more figuring is required.

Rules—A. Add together the angle of impulse and lever's angle of motion, then divide their sum by 2 and by means of the tables above mentioned find the sine of the answer.

B. Divide the angle of impulse by 2 and, as before, find the sine of the answer.

C. Divide the sine of the larger angle (Sub Rule B) by the sine of the lesser angle (Sub Rule A).

D. The quotient obtained by means of Sub Rule C should be divided into 1. The answer will be the modulus for the lever's acting length.

E. The above modulus multiplied by whatever the distance of centers may be will give the acting length of the lever suited to the center distance.

F. The length of the lever as calculated by Sub Rule E should next be multiplied by number of degrees representing the lever's angular motion and divided by the number of degrees contained in the roller jewel's impulse angle. The answer obtained will be the theoretical radius of the roller jewel.

G. Multiply the theoretical radius by .05 or .08 (sometimes more).

H. Add the product obtained according to Sub Rule G to the theoretical radius and answer will be the *Practical* radius of the roller jewel.

Example—The angular motion of the lever is 10 degrees. The angle of impulse is 30 degrees. The distance of centers is 10 millimeters. Calculate from above data the *practical* radius of the roller jewel. The practical radius being 5 per cent. longer than the theoretical.

Rules—A. $10° + 30° = 40° \div 2 = 20°$
Sine of $20° = .34202$
Sine of $15° = .25882$
B. $30° \div 2 = 15°$
C. $.34202 \div .25882 = 1.3215$
D. $1.00 \div 1.3215 = .7567$ lever modulus
E. $.7567 \times 10. = 7.56$ lever's length
F. $7.56 \times 10° \div 30° = 2.52$ theoretical radius
G. $2.52 \times .05 = .126$
H. $2.52 + .126 = 2.64$ practical radius

According to Sub Rule E the lever's acting length is 7.56 millimeters and the practical radius as shown by Sub Rule H is 2.64 millimeters. As previously mentioned, the length of the practical radius is a varying factor. It is somewhat dependent upon the specifications especially as regards the division of the total lock.

The following is taken from a large escapement drawing:
Distance of centers, 290 millimeters.
Impulse angle, roller jewel, 35°.
Angular motion of lever, 10°.

By means of the above, calculate the lever's acting length and practical roller jewel radius, allowing the practical an increase of 7 per cent. over the theoretical radius.

$$35^\circ + 10^\circ = 45^\circ$$
$$45^\circ \div 2 = 22\tfrac{1}{2}^\circ$$
$$\text{Sine of } 22\tfrac{1}{2}^\circ - .3827$$
$$35^\circ \div 2 = 17\ 1\text{-}2$$
$$\text{Sine of } 17\tfrac{1}{2}^\circ - .3007$$
$$38027 \div .3007 = 1.2727$$
$$1. \div 1.2727 = 227.82 \text{ mm. lever's length}$$
$$227.82 \times 10^\circ \div 35^\circ = 65.08 \text{ mm. theoretical radius}$$
$$65.08 \times .07 = 4.55$$
$$65.08 + 4.55 = 69.63 \text{ mm. practical radius}$$

The methods outlined for calculating lever's length and radius of roller jewel can be applied to advantage in checking drawings and models of the escapement. The calculations made also show that for different angles the modulus varies, therefore the modulus connected with the angles must be known before the lever's length can be determined.

LESSON 123

FACTS PRACTICAL AND THEORETICAL NO. 6

421. *The Division of the Lifts, With Regard to the Widths of Pallet and Tooth—Calculating Tooth's Freedom from Pallet Center.*—A few words about the divisional relationship existing between the width and lifts will be heplful to such as are interested in escapement drafting and construction of escapement models. As the amount of lift on a tooth, when the pallets are planted at the meeting points of the tangents is a corelated subject we shall also briefly discuss same.

No hard-and-fast rule can be followed for the relative proportions of the lifts to the widths, as a student will learn by investigating the various escapement specifications which come before him. A good general rule is to divide the lift as we divide the width subject to some modifications; for instance, making the tooth three-quarters the width of the pallet. For example, if the combined width of tooth and pallet, measured from the escape wheel center is 10½ degrees.

We can divide it as follows:

Width of pallet, 6°
Width of tooth, 4½°

On this basis of the division of the *width* we shall calculate the division of the *lifts*. Commence by expressing the foregoing widths in minutes:

Total width, 10½° = 630′
Pallet width, 6° = 360′
Tooth width, 4½° = 270′

We next figure what per cent. of the total width belongs to the tooth. This part of the problem can be solved by proportion as follows:

630 : 270 :: 100 ans. 42.8

This is practically 43 per cent. Therefore, 43 per cent. of the total width belongs to the tooth.

The next part of the problem is the *lifts*. The total combined lifts of tooth and pallet equals 8½ degrees. Changing 8½ degrees to minutes gives 510 minutes. As 43 per cent. of the *width* belongs to the tooth, then about 43 per cent. in the example we are figuring on belongs to the *lift*. Therefore 510. × 43. = 219.3 minutes or 3° 39′ represents the lift for the teeth. This is usually modified to some extent. In this particular instance we shall deduct 11 per cent. from the lift on the tooth and add it to the lift on the pallet—viz., 11 per cent. of 219.3 minutes equals 24., therefore 219. — 24. = 195′, or 3° 15′, which is the lift we assign to the tooth.

The total lift amounted to 8° 30′, subtracting the tooth's lift, 3° 15′ from this leaves 5° 15′ as the lift on pallet.

Here are the results tabulated:

Width tooth,	4½°
Width pallet,	6°
Lift tooth,	3¼°
Lift pallet,	5¼°

Calculations made with other escapement specifications will show various ways of dividing the width; and variations in the amount of the lift deducted from tooth and added to the pallet.

LESSON 124

FACTS PRACTICAL AND THEORETICAL NO. 7

422. *Freedom of Tooth's Heel from Pallet Center*—The below tables will be found useful for calculating the space existing between the heel of a tooth and the pallet center.

Pallets of the circular or equidistant type have, or should have, the pallet staff planted at the meeting point of the tangents, unless the escapement is very small. Small escapements if planted on tangents, owing to the lift on tooth, lack room between the heels of the teeth and pallet center; therefore but little space is left for the pallet staff. In larger escapements this difficulty is not experienced.

Moduli for Freedom of Pallet Center from Point of Tooth—

Lift tooth	Moduli
2¾°	.110
3°	.108
3¼°	.105
3½°	.103
3¾°	.101
4°	.100

Example—If the distance between the escape wheel and pallet centers measures 3.6 millimeters and the lift on tooth is 3 degrees; calculate by means of the above table the space separating the heel of tooth from pallet center:

Rule—Multiply the distance of centers by the modulus associated with the lift on tooth. Therefore:

$$3.6 \times .108 = .38$$

The answer, .38 millimeters, represents the space between heel of tooth and pallet center. This allows sufficient room for the pallet staff.

Alterations

ALTERATIONS NO. 1

423. *Remarks*—Scattered through this book are suggestions on the alteration of parts. In this series on alterations we have assembled explanations which beginners will find useful. The "don'ts" connected with alterations are here omitted.

424. *Diamond Laps*—Students interested in escapement work will find a set of diamond laps of different degrees of fineness a valuable acquisition, and are advised to either make or purchase same.

425. *Pallet Stone Setters*—For altering the lock on a pallet stone, use a tool so constructed that either stone can be heated independently of the other. A pallet stone setter of this class will save both time and trouble.

LESSON 125

ALTERATIONS NO. 2

426. *Grinding a Pallet Stone Thinner.—Method A*—Cement pallet jewel to the flattened end of a brass wire. The part of the stone to grind off is its back. The pallet jewel therefore should be cemented on the end of the wire with its back uppermost.

Place a diamond lap in the lathe. Revolve it at a moderate speed. Hold the pallet stone against the lap. Use a little oil to assist the grinding. A few minutes will complete the operation.

Method B—A wedged-shaped slice from the back of a pallet jewel near the releasing corner can be ground away, without the necessity of removing the pallet stone from its seat in the pallet arm. Methods A and B will prove useful for increasing the amount of drop or shake in lower grade watches.

LESSON 126

ALTERATIONS NO. 3

427. *Changing Angle of Lift on Pallet Jewel.*—When the lifting or impulse face of a pallet stone and the lifting face of a tooth show irregularities in the action of their lifts, we must, to correct the mismatched lifting action make some of the following alterations.

Method A—Supply a new pallet stone with a lifting face matching the lift on tooth.

Method B—Change the slant of the pallet stone in its seat so as to obtain a different lifting effect. This alters the relation of the pallet's impulse face with tooth's.

Method C—Sometimes to alter the slant of a pallet stone it is necessary to cut, by filing the sides of the seat in the pallet arms which retain the pallet jewel. This sometimes allows us to so pitch the stone that the matching is much improved.

Method D—To grind and polish to a different angle a pallet jewel's lifting face is a feat beyond the average horological mechanic. Hence there is no advantage in further discussing same.

LESSON 127

ALTERATIONS NO. 4 A

428. *Increasing a Lever's Entire Length.—Method A*—Supply a new and longer lever.

Method B—If the lever is soft it can be stretched by tapping with a hammer that part of the lever between the pallet staff and guard pin.

ALTERATIONS NO. 4 B

429. *Increasing a Lever's Acting Length.—Method A*—Supply a new lever.

Method B—Stretch the side walls of the slot by hammering, after which carefully redress and refinish the fork.

Method C—Advance the position of the roller jewel or supply a new table, one holding the roller more forward. Method C is an indirect way of correcting errors attributable to a short lever.

ALTERATIONS NO. 4 C

430. *Decreasing a Lever's Acting Length.—Method A*—Furnish a new lever.

Method B—File or grind away the horns, then refinish and polish them.

Method C—Errors caused by a long lever can at times be overcome by setting the roller jewel further back—that is, toward center of table.

LESSON 128

ALTERATIONS NO. 5 A

431. *Advancing the Position of the Guard Point.*—The intention of advancing the position of a guard point is to lessen the

freedom between guard point and table—viz., to decrease the guard freedom.

Method A—Supply a new table greater in diameter.

Method B—Remove old guard pin broach out of hole and insert a thicker pin.

Method C—Continue use of old pin, but make it "question mark" in shape.

Method D—If the guard point of a double roller escapement is short it can be made longer by squeezing with a pair of flat-nosed plyers, whose jaws inside are highly polished.

ALTERATION NO. 5 B

432. *When the Guard Point is Too Close.*—Should tests develop the fact that a guard point is too close to edge of table, their distance can be increased by using some of the following methods:

Method A—Select a new table lesser in diameter.

Method B—Place old table in the lathe. (Not infrequently it can be left on its staff. It is always advisable to remove the roller jewel.) Then, with either lap or graver, cut edge of table away. Of course a lap is preferable for this purpose. Afterward highly polish the table's edge.

Method C—Replace old guard pin with one that tapers slightly.

Method D—Use tool described in Lesson 134. Shave face and sides of the original guard pin.

Method E—If the guard point in a double roller is too long, use a fine file to shorten it, then burnish. (Note—If guard finger is short, grasp it with plyers; a slight squeeze will stretch it the desired amount.)

LESSON 129

ALTERATIONS NO. 6 A

433. *When the Roller Jewel and Slot Corners Are Too Close.*—*Method A*—Select a new table with roller jewel nearer center of table.

Method B—Enlarge hole in old table and set roller jewel further back.

Method C—Reduce the lever's acting length.

ALTERATIONS NO. 6 B

434. *When the Roller Jewel and Slot Corners Are Too Far Apart.*—*Method A*—Choose a new table, one holding the roller jewel more forward.

Method B—Enlarge hole in old table in such a way that the roller jewel will occupy a more advanced position.

Method C—Increase the lever's acting length.

LESSON 130

ALTERATIONS NO. 7 A

435. *When the Table is Too Large.—Method A*—Furnish a new table lesser in diameter.

Method B—Secure the table to the lathe; turn or grind off some of the edge. When finished the edge of table should show a high polish.

Method C—Frequently the old table can be used without altering it. Under such circumstances the guard pin can be made thinner, or the guard finger shortened.

ALTERATIONS NO. 7 B

436. *When the Table Is Too Small.—Method A*—Procure a larger table.

Method B—If the small table is retained advance the guard pin, making same "question mark" in shape. In double roller escapements the guard finger can be stretched in the manner directed in Lesson 128.

LESSON 131

ALTERATIONS NO. 8 A

437. *Increasing Drop and Shake.—Method A*—Employ thinner pallet jewels.

Method B—Grind off the entire back of one or both pallet stones.

Method C—Grind a V shaped slice off the back of pallet stone, the base of cut being the releasing corner.

Method D—Spreading both pallet stones apart increases inside drop and shake.

Method E—Closing both pallet stone—*i. e.*, bringing their respective releasing corners closer together increases outside drop and shake.

Method F—Increasing the drop lock by pushing out the receiving pallet increases inside drop and shake.

Method G—Increasing drop lock by pushing out the discharging pallet increases outside drop and shake. (The contrary is true of F and G when the stones are pushed back.)

NOTE—When either methods, D or E, are used, it is sometimes necessary to use a file to enlarge the slot containing each pallet stone.

NOTE—Increasing drop lock by pushing out one pallet stone increases the extent of drop lock on opposite stone. This the student can demonstrate by actual experiment.

ALTERATIONS NO. 8 B

438. *Lessening Drop and Shake.—Method A*—Use thicker pallet stones.

Method B—Spreading the stones apart decreases outside drop and shake.

Method C—Closing the stones together—that is, bringing their ends closer together—decreases inside drop and shake.

Method D—Increasing drop lock by pushing out the receiving pallet stone decreases outside drop and shake.

Method E—Increasing drop lock by pushing out the discharging pallet jewel decreases inside drop and shake.

NOTE—Lessening the drop lock on receiving pallet, by pushing it back in its seat, decreases inside drop and shake. Compare with Method F, Paragraph 437.

Lessening drop lock on discharging pallet, by pushing same back, decreases the outside shake. Compare with Method G, paragraph 437.

LESSON 132

ALTERATIONS NO. 9

439. *Improving Draw.—Method A*—As defective draw is sometimes due to unsuitable pallet stones, replace old with correct jewels.

Method B—Change the slant of offending stone in the desired direction. *Usually* increasing the pitch of a stone increases the draw, and *vice versa*, decreasing the slant, lessens draw. Extremes in pitch of stones must be avoided, as it destroys draw.

Method C—When an escape wheel shows signs of abuse the draw will be found irregular; some teeth will show good draw, others will not. Such a wheel should be replaced by a new one.

LESSON 133

ALTERATIONS NO. 10

440. *Straightening or Bending Levers.*—As a rule, thin levers, such as are found in 0-12 and 16 size watches, can be bent without any great risk of breakage.

Bending Tool, Method A—An excellent tool for bending levers can be made from plyers having one nose convex, the other concave. A set screw being so placed in the handle that the amount of bending is under the control of the set screw. Several sizes of tools are necessary to meet the varying lengths of levers.

Method B—Assuming the lever is in position in the watch

and as the workman looks at it the lever is at rest against the right-hand bank—first, take a fine screwdriver and, with the left hand, hold it against the left side of the lever. The point of the tool being back as far as possible, and placed under the bridge if it can be managed; second, take another but larger screwdriver, place it against the right side of the lever and with it bear against side of lever. In this way the lever can readily be bent. A few experiments will soon teach the operation.

LESSON 134

ALTERATIONS NO. 11

441. *Tool for Thinning Guard Pins.*—Select a needle or piece of steel wire the thickness of a fine darning needle. Drill a hole in one end, using a drill the size of a guard pin. The wire still being in the lathe, turn drilled end tapering. Then with an oil-stone slip sharpen the end. Next with a file flatten off one side, cutting away about one-third of the wire in order to reach the hole. Cutting into the hole drilled in the tool is done for clearance, so that when the tool is used—namely, slipped over the guard pin—brass shaving cannot become clogged inside. When finished the cutter should be tempered. A set of three having various sized holes to fit different thickness of guard pins will be found useful not only for thinning guard pins, but for removing the belly left on a guard pin when it is slightly inclined away from edge of table.

LESSON 135

ALTERATIONS NO. 12

442. *Guard Pins Shaped "Question Mark."*—When necessary to bring the guard pin more forward, it is best to shape it into a question-mark pin, such as is found in Waltham watches. This is an excellent shape, because of the latitude it allows for adjustment. A suitable tool can readily be made or purchased for forming pins into this shape.

QUESTIONS FOR RESEARCH WORK

The following list of questions, numbered consecutively from 1 to 418, has been compiled so that the important points involved in Escapement Knowledge may be brought before the student in detail. With each question is given the number of the paragraph in which an answer or explanation may be found. Research along these lines cannot fail to result in the student rapidly acquiring a thorough working knowledge of the problems relating to the subject upon which this work has been published.

Draw

443. *Questions on Draw.*—

1

Explain the term "draw." 23.

2

Why is the locking face of a pallet jewel given a slant? 110.

3

From what point does the draft angle of a pallet stone arise? 114, 214.

4

What is the cause of draw? 147.

5

In what way does the lever show the presence of draw? 147.

6

Is slide the result of draw? 148.

7

If the draw is poor will slide aid? 148.

8

When draw is defective, will the lever rest securely against its bank? 148.

9

How would you examine the draw? 152, 153, 154, 155.

10

What force aims to prevent contact of guard point with edge of table? 149.

11

If "draw" is not strong enough to retain the lever against its bank, will it cause trouble, and why? 149.

12

Is there any connection between slide and draw? 148.

13

If an escapement is banked to drop, will draw be present? 179.

14

Name the three phases of escapement action in a double roller escapement where draw must be effective. 151.

15

Name the three phases of action in a single roller escapement where draw must be effective. 150.

16

In a double roller escapement, when the guard finger is brought into contact with the edge of the table, what force releases them? 162.

17

Should a watch receive a shock when the guard finger enters the crescent and the lever thereby be jolted away from its bank what causes the lever to return to its bank? 153.

18

Should the slot corner be thrown in contact with the roller jewel will the parts mentioned remain in contact if the draw is sound? 154.

19

If a watch is clean and freshly oiled, and the draw is poor, how can the force of draw be increased? 156, 439.

20

When the guard pin is outside the crescent and the watch receives a violent shock sufficient to throw the lever off its bank, what parts are liable to come into contact? Also, will they remain in contact? 150.

21

Should the lever be thrown off its bank when the guard pin enters the crescent, name the parts which will come in contact, and state by what agency the parts are separated? 150.

22

When, from any cause, the slot corner and roller jewel are brought in touch with each other, name the force which returns the lever to its bank? 150.

23

By what means may the amount of draw be increased or decreased? 156, 439.

24

Suppose, to alter "draw" it becomes necessary to change the slant of a pallet stone, name the four points calling for investigation. 156.

Drop

444. *Questions on Drop.*—

25

What is drop? 17, 157.

26

State source of angle of drop and give its usual size. 127.

27

From what center is the angle of drop measured? 182, 213.

28

Into how many classes is drop divided? 128.

29

What is meant by "inside drop"? 18.

30

Define "outside drop." 19.

31

State whether the drops are more equal when steel escape wheels are used, and why. 159.

32

Explain the manner of testing inside drop. 163, 285.

33

How would you examine the outside drop? 161, 284.

34

Explain method whereby the extent of the angle of drop in an escapement can be approximated. 182, 184.

35

Which is least in amount, "drop" or "shake"? 158.

36

When, in order to correct an error of drop or of shake, we desire to alter but one pallet stone, how would you decide which stone to alter? 165, 437, 438.

37

When drop, or shake, is tight outside, how can it be corrected? 166, 437, 438.

38

If defective drop or shake is due to a thick pallet jewel, how would you remedy it? 167, 437.

39

Explain how you would correct drop or shake when tight inside. 165, 437, 438.

40

What is meant by the terms "outside" and "inside" as applied to the subjects of "drop" and "shake"? 18, 19, 20, 21.

41

If the amount of drop in an escapement equals one-half the width of the pallet jewel, and we assume the width of the pallet as 5 degrees, what size is the angle of drop? 182, 184.

Shake

445. *Questions on Shake.*—

42

Define the term "shake." 20.

43

Explain what the term "inside shake" means. 21.

44

What is meant by "outside shake"? 22.

45

When the drops are unequal, how will it affect the shakes? 130.

46

Can shake exist without drop? 130.

47

How many classes of shake are there? 130, 158.

48

How would you find out if shake is prseent? 21, 22.

49

Explain how you would test the "outside shake"? 162, 286.

50

State how the "inside shake" is tested? 164, 285.

51

Will want of shake cause a watch to stop? 130.

52

When irregularities in the shakes are discovered, what should be first examined? 165.

53

What important points require attention when shake is altered? 156.

54

When inside drop and inside shake is deficient, how would you provide same? 167, 437, 438.

55

Suppose we desire to change but one pallet stone to help correct an error of shake or drop, how would you decide which pallet stone to alter? 167, 437, 438.

56

When outside shake is wanting, how would you provide same? 168, 437, 438.

57

Given an escapement, in what way would you approximate the degrees of inside and outside shake? 184 A.

58

If the quantity of shake equals one-fifth the width of a pallet, state in degrees the approximate amount of shake present. 184 A.

The Lever

446. *Questions on the Lever.*

59

Define "lever." 47.

60

Define "horns of lever." 49.

61

Define "slot or notch." 50.

62

Define "fork." 51.

63

What parts of the lever comprise the fork? 135.

64

Where are the slot corners located? 135.

65

What is meant by the term "lever's acting length"? 48.

66

Explain what is meant by "run of lever"? 52.

67

Is it necessary for the lever of a single roller escapement to have long horns? 136.

68

Are the horns in a single roller escapement factors in the safety action? 136.

69

What is the purpose of the slot? 137.

70

Name and locate the lifting or impulse planes which move the lever? 137.

71

State where the angles relating to the fork originate and describe them. 138.

72

In a double roller escapement, how would you find out if the length of horn is correct? 207.

73

What force retains the lever against its bank? 149.

74

Are the slot corners factors in the safety action? 187, 199.

75

Is the acting length of the lever related in any way to the amount of drop lock? 269, 270.

76

Given the drop locks as correct, how would you decide if the lever's acting length is correct? 250, 254, 257.

The Pallets

77

What parts constitute the pallets? 53.

78

Name that part of the pallets which holds the pallet jewels? 54

79

Define "entering or receiving pallet." 57.

80

Define "exit or discharging pallet." 58.

81

Make a sketch and mark out a pallet jewel's locking and impulse face. 59, 61, 110.

82

Define "releasing corner of pallet." 62.

83

Explain why the locking face of a pallet jewel is given a slant. 110.

84

Explain purpose of the impulse plane on a pallet stone. 110.

85

Name, and state sources of all angles, giving shape to a pallet jewel. 111, 112, 113, 114.

86

Where does the angle of impulse which forms the lifting plane of a pallet arise? 112.

87

Name, and give origin of angle which controls the width of a pallet jewel. 113.

88

From what point does the draft angle of a pallet jewel arise? 114.

89

From what center does the angle of lock originate? 212.

90

Has the angle of lock any connection with the shape of the pallet jewel? 115.

91

Mention the important points which require attention when the position of a pallet jewel is altered. 156.

92

When an escapement trips, upon what part of the surface of a pallet jewel will the toe of the escape wheel tooth be found? 86, 192.

93

In American watches, which type of pallet is used—viz., circular or equidistant? 116.

94

How woud you recognize pallets of the circular type? 116.

95

Are the locking faces of circular pallets at an equal distance from the pallet center? 116.

96

How are pallets of the equidistant type recognized? Also, state if there are any inequalities in the measure of the distance between each locking corner and the pallet center? 117.

97

Should the complete pallets of a foreign make of watch become lost, how would you estimate the dimensions of new pallets? 414.

Lift on Pallet

448. *Questions—Lift on Pallet.—*

98

Locate the impulse face of a pallet jewel. 60, 110.

99

What is meant by "lift on pallet"? 60, 112.

100

Is the impulse plane related to the lift? 110.

101

The ange of impulse of a pallet stone arises from what center? 112, 212.

102

When the lifting planes of tooth and pallet start action, which of the following would you say is correct: (a) The pallet corner to start action on the tooth's impulse face, or (b) the toe of the tooth to commence action on the impulse face of the pallet? 170.

103

When the "lifts" are defective, what may be expected about the going of the watch? 170.

104

Explain how errors due to mismatched lifts can be lessened. 171.

Tooth of Escape Wheel

449. *Questions—Tooth of Escape Wheel.—*

105

Name the acting parts of a club tooth. 120.

106

What part of a club tooth should rest on the locking face of a pallet? 120.

107

Give location of tooth's impulse plane. 121.

108

Why is the line B. C., Fig. 9, of a tooth given a slant? 122.

109

Where does the lifting angle of a tooth originate? 124.

110

The degrees of width granted a tooth are measured from what center? 125.

111

From what point is the slant of a tooth's locking face measured? 126.

112

The shape of a club tooth is controlled by what angles? 123.

113

What feature governs the undercutting A to N, Fig. 9, which helps shape a tooth? 120.

114

If a ratchet tooth escape wheel or a club tooth escape wheel are destroyed or unfindable, explain how you would find the size of a new wheel. 413.

Lift on Tooth

450. *Questions—Lift on Tooth.—*

115

Explain the term "lift on tooth." 32.

116

Where is a tooth's impulse face located? 34, 121.

117

Locate the heel of a tooth. 35.

118

Locate the toe of a tooth. 36.

119

Why does a club tooth possess an impulse plane? 121.

120

Describe the correct relation of a tooth's lifting plane when acting upon the lifting plane of a pallet jewel. 169.

121

Where does the lifting angle of a tooth arise? 124.

Drop Lock

451. *Questions on Drop Lock.*—

122

From what center does the angle of lock arise? 212.

122 A

Define and explain the term "drop lock." 38, 173.

123

What controls the amount of drop lock? 173.

124

Is drop lock a product of the banking pins? 173.

125

As regards the drop lock, what is meant when we say an escapement is "banked to drop"? 173.

126

Is the amount of drop lock associated in any way with the safe action of an escapement? 217.

127

Take a movement and *estimate* the amount of its drop lock. 183.

128

How can the degrees of drop lock be measured? 417.

129

What does the expression "correct drop lock" mean? 268, 269, 270.

130

Would the extent of drop lock exactly suited to a high grade watch be equally well adapted to a low grade watch? 265.

131

On the basis of drop lock, name the three divisions into which for practical reasons escapements may be separated. 266.

132

What is meant by "a perfect escapement"? 268.

133

What does the term "correct escapement" imply? 269.

134

Explain what is meant by the expression "commercially correct escapement." 270.

135

Has the amount of drop lock any relationship to the lever's acting length? 250.

136

Which should be greatest, guard freedom or drop lock? 322.

137

In an Elgin type of escapement, which should be least, corner freedom or drop lock? 322

138

Does the extent of drop lock in an Elgin type of escapement differ from that found in an escapement of the South Bend type? 275, 276.

138, Section 1

If the receiving stone is pushed out, thereby making its drop lock greater, how will it affect the following: (a) Drop lock on discharging pallet, (b) inside drop, (c) inside shake? 308 (note).

138, Section 2

When the discharging stone is pushed out, thereby making its drop lock greater, how are the following affected: (a) Drop lock on receiving stone, (b) outside drop, (c) outside shake? 308 (note).

138, Section 3

Suppose we lessen the drop lock on the receiving stone by pushing this stone back into its seat, what effect will this have on the following: (a) Drop lock on discharging pallet, (b) inside drop, (c) inside shake? 309.

138, Section 4

By pushing the discharging stone back its drop lock is decreased; mention the effect this will have on the following: (a) Drop lock on receiving pallet, (b) outside drop, (c) outside shake. 309.

Slide

452. *Questions on Slide.*—

139

Define "slide." 41

140

What is meant by slide or slide lock? 148, 174.

141

What controls the amount of slide? 148, 174.

142

Does any relationship exist between "slide" and "draw"? 148.

143

If an escapement is banked to drop, would slide be present? 174.

144

Is the run of the lever related to slide? 52.

145

What is meant by "banked for slide"? 16.

146

How can the slide lock be increased? 174.

147

Is the amount of guard freedom and amount of corner freedom related to slide, and in what way? 244 to 248.

Total Lock

453. *Questions on Total Lock.*—

148

Define "total lock." 45.

149

Of what is the total lock composed? 172.

150

Given an escapement, how would you approximate the degrees of total lock? 183.

151

How can the degrees of total lock be measured? 417.

Safety Lock

454. *Questions on the Safety Lock.*—

152

Define "safety lock." 44.

153

Explain the purpose of a safety or remaining lock. 176.

154

Name the three safety locks. 402.

155

When tests show an absence of safety lock, what error develops? 176.

156

How would you demonstrate the presence of a guard safety lock, a corner safety lock, a curve safety lock? 193, 194, 195, 204, 205, 206.

Bank and Banking Pins

455. *Questions—Bank and Banking Pins.*—

157

Explain the term "bank." 12.

158

Define "banking pin." 11.

159

Describe the banking pins and their purpose. 178.

160

What is expressed by the term "banking error." 10.

Overbanking

456. *Questions on Overbanking.*—

161

Define "overbanked." 46.

162

What is meant when we say "an escapement is overbanked"? 191.

163

Name some causes of overbanking. 191.

Banked to Drop

457. *Questions—Banked to Drop.—*

164

What does the term "banked to drop" imply? 13.

165

Describe the best method for unlocking escapement errors. 179.

166

What is meant when we say "a watch is banked to drop"? 179.

167

Explain the method of banking an escapement to drop. 223.

168

When an escapement is banked to drop will slide be present? 179.

169

Is draw present when an escapement is banked to drop? 179.

170

When a watch of the Elgin type is banked to drop, will freedom be found between the guard point and the table? 14, 179, 288.

171

If an Elgin type of escapement is banked to drop, will corner freedom be present? 14, 293.

172

Is it correct or incorrect to find guard freedom present when a Dueber or South Bend escapement is banked to drop? 15, 291.

173

When a watch of the South Bend type is banked to drop, will corner freedom be present? 15, 296.

174

When banked to drop and escapement parts are well matched, state the proof-findings of the following guard, corner and angular tests, the watch being of the Elgin type? 15, 291.

175

In a South Bend type of escapement, when the parts are well matched and watch is banked to drop, what are the proof test findings of the guard, corner and angular tests?

Impulse and Safety Tables

458. *Questions—Impulse and Safety Tables.—*

176

Describe the roller table found in single roller escapements. 76, 144.

177

Describe a safety roller. 74.

178

What is meant by diameter of table? 76.

179

What is the necessity for a crescent or passing hollow? 75, 146.

180

In a double roller escapement, what table is associated with the safety action? 144.

181

When the guard point is thrown in contact with the edge of the table, what force causes the parts to separate? 150, 151.

182

From what center does the angle originate, which provides freedom between edge of table and guard pin? 211.

183

Under "banked to drop" conditions, will the guard point touch the edge of a table (a) in an Elgin, (b) in a Dueber? 14, 15, 226, 228.

Roller Jewel

459. *Questions—Roller Jewel.—*

184

By what other names is the roller jewel known? 65.

185

Why is the face of a roller jewel flattened? 141.

186

What is meant by "roller jewel radius"? 66.

187

Define "impulse radius." 143.

188

From what center is the width of a roller jewel measured? 143.

189

Explain how you would find out if the roller jewel fits the lever slot. 287.

190

Describe how a roller jewel should be reset so that its position matches the escapement action (a) in an Elgin, (b) in a South Bend escapement. 267, 271, 272.

191

Describe the various positions and actions of a roller jewel when a watch is running. 142.

192

Elgin, B to D—where does the angle arise which provides freedom between the roller jewel and slot corner? 143.

193

Does slide increase the corner freedom? 143.

194

If an escapement of the Elgin type is banked to drop, would you expect to find freedom between the slot corner and roller jewel? 14, 240.

195

Name the escapement type, which, when banked to drop, allows no freedom between slot corners and roller jewel. 240.

196

When by accident the slot corner is thrown into contact with the face of the roller jewel, slide being present, what force pulls them apart? 148.

197

The roller jewel as a part factor in the safety action is associated with what parts? 187, 199.

198

In a single roller escapement, what are the functions of the roller jewel as a factor in the safety action? 189.

199

In a double roller escapement, when the lever horn is thrown in contact with the roller jewel, what causes the release of the parts? 148.

200

Describe the work of the roller jewel as a factor in the safety action of a double roller escapement. 201.

201

Explain relation of the lever horns to the roller jewel in single roller escapements. 196, 207, 298, 300.

202

Explain the curve test and its purpose. 197, 208.

Guard Point

460. *Questions on the Guard Point.*—

203

Locate the guard pin and guard finger. 68, 69.

204

What is meant by "guard radius"? 71.

205

Name source of angle which provides freedom between guard point and table. 145.

206

When the lever is at rest against its bank, slide being present, what force guards against contact of guard point with table? 147.

207

State three important positions of the guard pin during the routine action of an escapement. 186.

208

The escapement being in action, when is the guard pin closest to the edge of the table? 186.

209

When is the guard pin at its greatest distance from edge of table? 186.

210

As a factor in the safety action, name the parts with which the guard pin is associated. 187.

211

Name the functions of the guard finger as it relates to the safety action. 188.

212

In a single roller escapement, when the guard pin just enters the crescent, in what position is the roller jewel? 185.

213

In a double roller escapement, when the guard finger enters the crescent, state what part of the fork the roller jewel is then opposite? 198.

214

Describe the functions of the guard finger in a double roller escapement. 188.

215

While the guard finger of a double roller escapement remains within the crescent, how is the safety action preserved? 201.

216

When the guard point is held in contact with the edge of the table, what should the effect be as regards the lock? 204.

Safety Actions, S. R.

461. *Questions on the Safety Actions, Single Roller Escapement.*—

217

What is the purpose of the safety actions? 187.

218

In single roller escapements, name the parts which insure the safe action of an escapement. 67, 187.

219

Describe the office of the guard pin as a factor in the safety action. 188, 200.

220

The roller jewel as a factor in the safety action is associated with what parts? 189.

221

What is meant by "overbanking"? 190.

222

What causes overbanking? 191.

223

Explain the term "remaining lock." 44.

224

Should a watch receive a jolt when the crescent is well past the guard pin, explain how the safe action of the escapement is preserved. 190.

225

In the event of the lever leaving its bank during the time the guard pin is within the crescent, what parts are then called upon to preserve the escapement from going out of action? 184

226

What insures the escapement remaining in action, should the lever be thrown away from its bank when the roller jewel is opposite the slot corner? 201.

227

When the roller jewel is opposite that part of the lever horn, *near* the slot corner, what preserves the safety action of the escapement should the lever be thrown off its bank? 188, 196, 197.

Safety Actions, D. R.

462. *Questions on the Safety Actions—Double Roller.—*

228

Name the safety action parts of a double roller escapement. 199.

229

What is the preventative function of the guard finger? 200.

230

Name the functions of the roller jewel as a part of the safety action. 201.

231

How would you decide if the length of horns are correct? 207.

232

While the guard finger remains outside the crescent, what parts when called upon insure the escapement remaining in action? 201.

233

As a part of the safety action, with what is the guard finger associated? 199.

234

The lever horn is associated with what part as a preserver of the safety action? 201.

235

Name the chief parts which act as a preventative of overbanking. 200.

236

State with what parts the roller jewel is associated to insure the safe action of an escapement. 201.

237

When the guard finger is within the crescent, upon what parts does the protection of the escapement's action depend? 201.

237 A

Is there any relationship between width of crescent and length of the horn? 198 (note), 207.

Tripping

238

463. *Questions on Tripping—Single Roller.*—

Explain what is meant when we say "an escapement trips." 192.

239

Define the following: "Guard trip," "corner trip," "curve trip." 87, 88, 89.

240

In how many positions can a trip occur? 192.

241

Explain the use of the guard safety test. 193.

242

Can an escapement trip on some teeth while other teeth possess a safety lock? 193.

243

Is the amount of guard freedom related to the locks? 224, 289.

243 A

State causes for (a) guard trip, (b) a corner trip, when all escapement pivots fit their respective holes. 193, 194.

244

Describe use of corner safety test and explain its purpose. 194.

245

How would you make use of the curve safety test and why? 195.

246

If tripping errors are not corrected, what will the result be? 192.

247

Has the amount of corner freedom any relation to the amount of drop lock? 225.

464-465. *Questions on Tripping—Double Roller.*—

248

Name the three positions wherein to suspect the presence of tripping errors. 203.

249

How would you make use of the guard safety test? and tell why. 204.

250

How, and for what purpose is the curve safety test used? 206.

251

The amount of lock and the amounts of guard and corner freedoms are related in both single and double roller escapements. State why the locks and freedoms are related. 225, 294.

Angular Test

466. *Questions on the Angular Test.*—

252

Describe the system best adapted by beginners for applying the angular test. 258.

253

Why is it advisable for beginners to remove the guard point from edge of table when using the angular test? 259.

254

Explain method of blocking the lever when making use of the angular test. 258.

255

When the lever is blocked, at what moment should we cease rotating the balance? 258.

256

If the locks are correct, and the lever's acting length is adapted to the locks, when the angular test is used, state what position each tooth will occupy on each pallet (Elgin). 258.

257

What do the "proof findings" of the angular test imply? 190.

258

Describe a variation from the angular test's proof findings. 258.

259

"Out of Angle" is shown in what manner by the angular test? 262.

260

What is the usual cause of an escapement being "out of angle"? 263.

261

Describe the angular test's proof findings for an escapement of the "South Bend type." 257.

262

In what way does the proof findings of an Elgin type of escapement differ from the proof-findings of a South Bend or Dueber escapement? 254, 257.

263

Granted that the drop locks in an escapement of the Elgin type are correct, how would you prove that the length of the lever is right? 258.

SUMMARY OF TESTS

The Summary Includes 468 to 473

467. *Questions on Summary of Tests.*—

264

Describe method for testing "draw." 281.

265

How would you test drop lock? 282.

266

Explain how you would test "drop," inside and outside. 283' 284.

267

Describe manner of testing the inside and outside shakes. 285, 286.

268

How would you determine the freedom of the roller jewel when held by the slot walls? 287.

The Guard Test

468. *Questions on the Guard Test—Single and Double Roller.—*

269

Explain the term "guard freedom." 84.

270

What is the nature and purpose of the guard test? 288.

271

When we bank to drop escapements of the Elgin and South Bend types, state in which type guard freedom would be present. 288.

272

Describe methods of making the guard test in escapements of the Elgin and South Bend types. 290, 291.

273

Explain the three classifications into which guard freedom may be divided. 304.

274

When the guard test is applied to a South Bend escapement, same being banked to drop, can the lever be lifted off its bank? Also, under like conditions, can the lever in an escapement of the Elgin type be lifted away from its bank? 290, 291.

The Corner Test

469. *Questions on the Corner Test—Single and Double Roller.*—

275

What is meant by the term "corner freedom"? 79.

276

Describe manner of testing the corner freedoms in escapements of the Elgin and South Bend types. 295, 296.

277

If a watch of the Elgin type is banked to drop, would you consider the escapement correct if corner freedom is *not* present? 293.

278

If we bank to drop escapements of the Elgin and South Bend types, would their corner freedoms be identical? 293, 295, 296.

279

Into how many types may corner freedom be divided? 303.

The Guard Safety Test

470. *Questions on the Guard Safety Test—Single and Double Roller.—*

280

What does the term "guard trip" imply?

281

For what purpose is the guard safety test employed? 289.

282

When using the guard safety test, is it necessary to bank the escapement drop? 289.

283

Describe routine of making the guard safety test. 292.

The Corner Safety Test

471. *Questions on the Corner Safety Test—Single and Double Roller.—*

284

What is meant by the expression "corner trip"? 87.

285

If the corner safety test showed that the safety lock is uncertain, or absent, would it require correction? 192, 194.

286

Describe manner of using the corner safety test. 297.

287

When the corner safety test is employed, is it essential that the escapement be banked to drop?

288

How would you determine the condition of the corner safety lock? 297.

Curve Test and Curve Safety Test, S. R.

472. *Questions on the Curve Test and Curve Safety Test—Single Roller.—*

289

Describe how the curve test is made, and its purpose. 298.

290

Should the curve test show that the roller jewel catches on the lever horns, would you consider the action correct? 298.

291

When using the curve test, what controls the extent of the horn with which the roller jewel can come in contact? 298.

292

What is the "curve safety test" and how is it employed? 299.

Curve Test and Curve Safety Test, D. R.

473. *Questions on the Curve Test and Curve Safety Test—Double Roller.—*

293

473. Describe method of using the curve test. 300.

294

When the curve test is used, state at what moment you would expect the roller jewel to come in contact with the horns of the lever. 300.

295

What is meant by "curve trip"? 88.

296

State the maner in which you apply the curve safety test. 301.

Tests and Escapement Testing

474. *Questions on the Tests and Escapement Testing.—*

297

Before testing an escapement, what points require attention? 278.

298

Describe the routine of an escapement examination. 279.

299

If we desire to learn the relation of a guard point with its table, what tests are used? 290, 291.

300

When we wish to investigate the relation of the roller jewel with the slot corners, what test is employed? 295, 296.

301

What test informs us about the relation of the roller jewel to the curves of the horns? 298.

302

To investigate the condition of the safety actions in single and double roller escapements, name the tests employed? 292, 297, 301.

303

What test informs us if the extent of drop lock present in any escapement is exactly adapted to the acting length of the lever? 250.

Escapement Types

475. *Questions—Escapement Types.*

304

475. How many types of escapement are used in American watches? 274.

305

Of the total lock in an escapement of the Elgin type, how much belongs to the drop lock and how much is slide? 275.

306

In South Bend escapements, how much of the total lock is slide and how much is drop lock? 276.

306 A

Suppose you desired to examine the condition of an escapement in a foreign built watch, which type of American escapement will nearest apply? 276 (note).

Rules and Alterations

476. *Questions on Rules and Alterations.—*

307

When the drop locks are increased—that is, *made deeper*—what effect will changing the locks have on the following: (a) The bankings, (b) the guard freedoms, (c) the corner freedoms, (d) the safety locks? Alteration, 308, A, B, C, D.

308

If we make the drop locks *lighter*, describe effects produced on (a) the bankings, (b) the guard freedoms, (c) the corner freedoms, (d) the safety locks. Alteration, 309, A, B, C, D.

309

Describe how the corner freedoms may be increased. Alterations Nos. 1 and 2, 310.

310

Explain what alterations will decrease corner freedoms. Alterations Nos. 1 and 2, 311.

311

Name the changes whereby guard freedom can be increased. Alterations 1 and 2, 312.

312

Describe alterations which will decrease guard freedoms. Alterations 1 and 2, 313.

313

Given two escapements, both of them alike and identical, to which the following alterations are made: (a) In one we decrease the lever's acting length, (b) in the other we increase the drop locks; (c) state, if after alterations, their corner freedoms are similarly affected; (d) how do their respective corner safety locks show the change? 310.

314

If in this instance we are given two escapements, both of them correct and alike in their details, (a) when we increase the lever's acting length in one; and (b) decrease the drop locks in the other, will (c) the alterations cause resemblance in their respective corner freedoms, (d) in what way will the corner safety locks of each reflect the alterations? 311.

315

Assuming we have two escapements exactly alike and correct, (a) one of these we alter by bending the guard pin away from edge of table, (b) the other escapement we change by making the drop locks deeper, (c) state how the guard freedoms in each resemble each other. (d) In what way will the alteration in each escapement affect the guard safety locks? 312.

316

Again, assuming we have two escapements alike correct and duplicates, (a) the first we alter by bending the guard pin closer to edge of table, (b) the second we change by decreasing the drop locks, (c) will their respective guard freedoms show any similarity? (d) Also, describe how the guard safety locks reflect the changes. 313.

317

When the position of the roller jewel is advanced, or the lever's acting length is made shorter, state results as regards the corner freedoms and corner safety locks. 314.

318

When the lever is cut, thereby making the lever's acting length shorter, what is the effect on (a) the corner freedoms, (b) the corner safety locks? 315.

319

When the bankings are spread apart, how will it affect the following: (a) The guard freedoms, (b) the corner freedoms, (c) the slide, (d) run of lever? 316.

320

State how the following are affected when the bankings are brought closer together: (a) The guard freedoms, (b) the corner freedoms, (c) the slide, (d) run of lever. 317.

321

When an Elgin type of escapement is banked to drop, will we find (a) any guard freedom, (b) any corner freedom? 318.

322

When an Elgin type of escapement is banked to drop, will the safety lock be equal or less than the drop lock? 318.

323

When banked to drop, will a South Bend type of escapement show any corner or guard freedom? 318.

324

Will the drop lock in a South Bend escapement equal or be less than the drop lock? 318.

325

What alterations must be made to correct a butting error? 319.

326

State what protects the safety lock. 322.

327

If an escapement is " out of angle," how will such a condition be expressed by the angular, corner, and guard tests? 323.

The Corner Test

477. *Questions on the Corner Test.—*

NOTE—When the abbreviations B. to D. are found opposite a question it indicates that the question expresses the condition of the escapement *when banked to drop.*

328

Elgin, B. to D.—Describe the *proof findings* of the corner test. 325.

329

Elgin, B. to D.—In this watch, which is of low grade, the drop locks are *unsafely light,* but the corner freedoms are *approximately correct.* The question is, when the drop locks are increased, in what way will it *alter* the corner freedoms, and what other changes might be required? 326.

330

Elgin, B. to D.—In this escapement the corner freedoms are seemingly *right,* but the drop locks are both *deep.* The problem is, if the drop locks are made correct—*that is, lighter*—how will this alteration act on the corner freedom, and what additional changes may be looked for? 327.

331

Elgin, B. to D.—In this instance the drop locks *are correct,* but the corner freedoms *are too great.* What changes are necessary to overcome the surplus corner freedoms? 328.

332

Elgin, B. to D.—When the drop locks are *rather light* and there is *too much* corner freedom, state how the corner freedoms will be affected, when the drop locks are made deeper, and what other alterations will be required to improve the escapement? 329.

333

Elgin, B. to D.—In this escapement we find the drop locks *are deep,* on testing the corner freedoms we find them *excessive.* What is the remedy? 330.

334

Elgin, B. to D.—The drop locks *are correct,* but by the corner test we find *an absence of* the corner freedoms. How can we overcome the defect? Also, state results. 331.

335

Elgin, B. to D.—When the drop locks are light and the roller jewel is unable to emerge from the slot, what alterations are necessary? 332.

336

Elgin, B. to D.—Assuming an escapement with the following troubles, how would you rectify the errors: Drop locks deep, roller jewel unable to make its exit out of the slot? 333.

337

Elgin, B. to D.—When there is an absence of the corner freedoms, and the drop locks are correct, what alterations would be called for? 334.

338

Elgin, B. to D.—We find that the drop locks are rather light and the roller jewel is unable to leave the slot. What changes are necessary in this escapement? Also, name the order in which alterations should be made? 335.

339

Elgin, B. to D.—The defects in this escapement are deep drop locks and an absence of all corner freedoms. Explain how you would proceed to improve the escapement? 336.

Corner Safety Test

478. *Questions on the Corner Safety Test.—*

340

Elgin, B. to D.—Given the drop locks as correct and the corner freedoms as likewise correct, but an examination shows that *some* teeth of the escape wheel will trip while others show a safety lock. Explain the cause of the tripping error and how a correction can be made? 337.

341

Elgin, B. to D.—An examination of this watch shows the drop locks as correct, but the corner freedoms are altogether too great. State what error will be found, and describe the alterations which will improve the escapement. 338.

342

Elgin, B. to D.—This escapement possesses the correct amount of corner freedoms, but the drop locks are decidedly too light, sufficiently so to cause tripping errors. How would you overcome the trip and what other changes might it be necessary to make? 339.

343

Elgin, B. to D.—Given the drop locks as deep, and the corner freedoms so excessive, they allow tripping to take place. In what order and way should alterations be made to restore the escapement to a more perfect condition? 340.

The Guard Test

479. *Questions on the Guard Test.—*

344

Elgin, B. to D.—Describe the proof findings of the guard test. 325.

345

Elgin, B. to D.—If the guard freedoms are right, but the drop locks are too light, what will be the effect on the escapement if the locks are increased? 341.

346

Elgin, B. to D.—When the drop locks are deep and the guard test shows the guard freedoms as practically correct, mention the changes involved after correcting the error in the locks? 342.

347

Elgin, B. to D.—This escapement's condition is as follows: Guard freedoms are slightly excessive, drop locks correct. State how would you alter the error in the guard freedoms. 343.

348

Elgin, B. to D.—The guard freedoms in this escapement are excessive, the drop locks are very light. What is the first alteration, and how does it affect the banking pins and the guard freedoms? Mention what other change is demanded. 344.

349

Elgin, B. to D.—Assuming an escapement with deep drop locks and a surplus amount of guard freedoms, state the required corrections. 345.

350

Elgin, B. to D.—In this escapement the drop locks are correct, but the guard test shows no freedom between the guard point and the table. What changes are implied? 346.

351

Elgin, B. to D.—When we discover an escapement wherein the drop locks are undoubtedly light, with no freedom between guard point and table, what is the first alteration? Also, state the additional changes which might be necessary. 347.

352

Elgin, B. to D.—Given the drop locks as deep and a condition of contact between guard point and table, what correction should be made at the start? And explain the nature of the alterations which follow? 348.

353

When the guard point butts or sticks against the edge of the table and no legitimate manipulation of the guard point will overcome the defect, how can the butting error be remedied? 349.

353 A

If when setting a watch, you push the second hand backward, and the watch stops, explain cause of error and its correction. 349.

The Guard Safety Test

480. *Questions on the Guard Safety Test.*—

354

Elgin, B. to D.—If the drop locks are quite right, and the guard freedoms are satisfactory, but we find that *some* teeth of the escape wheel trip while the majority show a safety lock, describe cause of trouble and how the tripping error can be remedied. 350.

355

Elgin, B. to D.—When we discover an escapement wherein the drop locks are correct, but the guard test reveals too much guard freedom, the result being a tripping error, explain how the escapement may be improved. 351

356

If a trip is discovered under the following circumstances, drop locks light, guard freedoms correct, in what manner would you overcome the tendency to trip? 352.

357

In this instance the drop locks are light and the guard freedoms excessive. What error would be present and what corrections must be made? 353.

358

What correction will be necessary when the drop locks are deep and the guard freedoms excessive? 354.

Angular Test and Out of Angle

481. *Questions on the Angular Test and out of Angle (see paragraph 459).*—

359

481. Assuming that the lever's acting length is correct, but the drop locks are *too light*, in what way will the angular test express the error of light locks? 358 (No. 4), 356.

360

Given the drop locks as *deep* and the lever's acting length as correct, explain how the angular test will show the error in the locks. 359 (No. 5), 356.

361

When the drop locks are *correct*, but the lever's acting length is too long, in what way will the excess in length of lever affect the teeth and pallets as determined by the angular test? 360 (No. 2), 356.

362

How will the angular test show the defect in the lever's acting length, when the lever is *short* and the drop locks *correct?* 361 (No. 3), 356.

363

What is meant by the term "out of angle"? 91.

364

When an escapement is out of angle, how will the defect be shown by the following: (1) The drop locks, (2) the relation of guard point with table (3) the relation of slot corners to roller jewel, (4) the relation of teeth and pallets as determined by the angular test? 363.

365

In this watch the lever is straight, but the drop locks we find are very unequal. Describe the necessary correction and how the defect in the locks is exposed by the angular, guard and corner tests. 364, 366, 368.

366

When we find the drop locks are unequal and besides the lever is bent, what is the first alteration, and what other changes are always necessary? 365, 367, 369.

Curve and Curve Safety Tests

482. *Questions on the Curve and Curve Safety Test—Double Roller (see paragraph 466).*

367

482. What is meant by curve test and curve safety test? 81, 82.

368

Should it be possible for the roller jewel to touch the end of the horn, how would you verify that contact of these parts is possible? 371.

369

How would you determine if the central part of the horn can seriously catch and hold the roller jewel? 372.

370

Explain method whereby you could find out if the roller jewel can stick, or catch on or near the slot corners. 373.

371

Describe the method of using, and nature of the curve safety test. 374.

483. *Questions on the Curve Safety Test—Single Roller (see paragraph 465).*

372

How would you prove that the roller jewel cannot touch the end or central part of the lever horn? 375.

373

State and describe the method, by means of which it can be learned if it is possible for the roller jewel to catch on either the slot corner, or that part of the horn near the slot corner. 376.

374

What is the curve safety test used for? And describe how it is used. 377.

Hints and Helps

484. *Questions on Hints and Helps.—*

375

If the lock is considered "rather light," explain how it can be determined if such is a fact. 401.

376

Name the three positions where the safety locks require attention and examination? 402.

377

The bankings in this watch have been tampered with, besides both of the pallet jewels have been taken out of their settings. Describe the guard and corner test methods for correctly resetting the pallet jewels and successful rematching of the escapement action? 403, 404, 405.

378

When one pallet is left in place, the opposite stone being removed from its seat, describe method used to secure a correct resetting of the loose pallet stone. 406.

379

Elgin, B. to D.—If you found contact of guard point with table, what changes and alterations can be made to improve the escapement and overcome the error? 407.

380

B. to D.—Should it be discovered that the guard point touches the edge of the table in some places, and is free in others, what would this indicate, and name remedies? 408.

381

B. to D.—How would you find out if the edge of the roller or safety table is running "true in the round"? 409.

382

Suppose you lost a roller table, state method of selecting a new one, right in diameter, and with the roller jewel correctly placed. 410.

383

The position the roller jewel occupies in a table is of great importance. If one is to be reset, state how it should be done to obtain correct rematching of its action. 411.

384

Describe the practical application of the corner test without having to bank the escapement to drop. 412.

385

If you lost an English ratchet tooth escape wheel, how may the correct size of a new one be determined? 413.

386

When an escape wheel of the club tooth form is lost, how would you find the dimensions of a new one? 413.

387

Take any watch and prove that its escape wheel is correct in size. 413.

388

Given an escapement model, or a drawing, state how you verify if the size of the escape wheel is correct. 413.

389

Should pallets belonging to an equidistant type of escapement become destroyed or lost, describe how you would go to work to select new and suitable pallets. 414.

390

If pallets belonging to the circular class be mislaid or lost, and the watch is of foreign make, how would you attempt the selection of suitable sized pallets? 414.

Theory and Practice

485. *Questions on Theory and Practice.*—

391

What is meant by "the angular motion of the lever"? 416.

392

What does the expression "roller jewel's impulse angle" imply? 416.

393

Explain how the lever's angular motion may be measured. 417.

394

Describe how degrees of lock can be measured.

395

How would you measure the degrees of lift on a tooth of an escape wheel having club teeth? 417.

396

State how the lift on a pallet stone can be measured. 417.

397

How would you measure the size of the impulse angle? 417.

398

Explain the proportional method for calculating a lever's acting length and the radius of the roller jewel? 418.

399

Is there any difference between the practical and the theoretical radius of the roller jewel? 419.

400

Given the degrees of lever's angular motion, and the impulse angle, how would you calculate the practical radius of the roller jewel? 420.

401

When "the lifts" are divided, is there any relationship between the amount of lift assigned to the tooth and amount given the pallet? 421.

402

Is it possible to plant all escapement on tangents? 422.

403

Give reason, why some pallets cannot be planted on tangents. 422.

Altering Parts

486. *Questions on Altering Parts.*—

404

Describe how a pallet stone may be ground thinner. 426.

405

How may the angle of lift on a pallet jewel be altered? 427.

406

In what way can the entire length of a lever be increased? 428.

407

Can the lever's acting length be increased, and how? 429.

408

How would you decrease the acting length of a lever? 430.

409

Explain the methods whereby the guard points' position could be advanced? 431.

410

When the guard point is "too close," state the remedies which may be used. 432.

411

When the roller jewel and slot corners are "too close," how should corrections be made? 433.

412

If the roller jewel and slot corners are too far apart, what alterations will overcome such a defect? 434.

413

Should the roller table be too large, describe changes which will correct the trouble? 435.

414

If a roller table is too small, what remedies would you suggest? 436.

415

Describe how "drop" or "shake" can be increased. 437.

416

If the "drop" or "shake" in an escapement is found to be too great, explain how same may be altered. 439.

417

Describe how "draw" may be improved. 439.

418

How would you lessen "draw"? 439.

INDEX TO SUBJECTS

A

B

C

D

E

H

I

L

O

M

P

Q

R

S

T

V

W

Made in United States
North Haven, CT
12 June 2023